OPEN DATING
OF
FOODS

F N P

PUBLICATIONS IN FOOD SCIENCE AND NUTRITION

Books

OPEN DATING OF FOODS, T.P. Labuza and L.M. Szybist
NITRITE CURING OF MEAT: *N*-NITROSAMINE PROBLEM, R.B. Pegg and F. Shahidi
DICTIONARY OF FLAVORS, D.A. DeRovira
FOOD SAFETY: THE IMPLICATIONS OF CHANGE, J.J. Sheridan *et al.*
FOOD FOR HEALTH IN THE PACIFIC RIM, J.R. Whitaker *et al.*
DAIRY FOODS SAFETY: 1995-1996, A COMPENDIUM, E.H. Marth
OLIVE OIL, SECOND EDITION, A.K. Kiritsakis
MULTIVARIATE DATA ANALYSIS, G.B. Dijksterhuis
NUTRACEUTICALS: DESIGNER FOODS III, P.A. Lachance
DESCRIPTIVE SENSORY ANALYSIS IN PRACTICE, M.C. Gacula, Jr.
APPETITE FOR LIFE: AN AUTOBIOGRAPHY, S.A. Goldblith
HACCP: MICROBIOLOGICAL SAFETY OF MEAT, J.J. Sheridan *et al.*
OF MICROBES AND MOLECULES: FOOD TECHNOLOGY AT M.I.T., S.A. Goldblith
MEAT PRESERVATION, R.G. Cassens
S.C. PRESCOTT, PIONEER FOOD TECHNOLOGIST, S.A. Goldblith
FOOD CONCEPTS AND PRODUCTS: JUST-IN-TIME DEVELOPMENT, H.R. Moskowitz
MICROWAVE FOODS: NEW PRODUCT DEVELOPMENT, R.V. Decareau
DESIGN AND ANALYSIS OF SENSORY OPTIMIZATION, M.C. Gacula, Jr.
NUTRIENT ADDITIONS TO FOOD, J.C. Bauernfeind and P.A. Lachance
NITRITE-CURED MEAT, R.G. Cassens
POTENTIAL FOR NUTRITIONAL MODULATION OF AGING, D.K. Ingram *et al.*
CONTROLLED/MODIFIED ATMOSPHERE/VACUUM PACKAGING, A.L. Brody
NUTRITIONAL STATUS ASSESSMENT OF THE INDIVIDUAL, G.E. Livingston
QUALITY ASSURANCE OF FOODS, J.E. Stauffer
SCIENCE OF MEAT & MEAT PRODUCTS, 3RD ED., J.F. Price and B.S. Schweigert
ROLE OF CHEMISTRY IN PROCESSED FOODS, O.R. Fennema *et al.*
NEW DIRECTIONS FOR PRODUCT TESTING OF FOODS, H.R. Moskowitz
PRODUCT DEVELOPMENT & DIETARY GUIDELINES, G.E. Livingston, *et al.*
SHELF-LIFE DATING OF FOODS, T.P. Labuza
ANTINUTRIENTS AND NATURAL TOXICANTS IN FOOD, R.L. Ory
POSTHARVEST BIOLOGY AND BIOTECHNOLOGY, H.O. Hultin and M. Milner

Journals

JOURNAL OF FOOD LIPIDS, F. Shahidi
JOURNAL OF RAPID METHODS AND AUTOMATION IN MICROBIOLOGY, D.Y.C. Fung and M.C. Goldschmidt
JOURNAL OF MUSCLE FOODS, N.G. Marriott and G.J. Flick, Jr.
JOURNAL OF SENSORY STUDIES, M.C. Gacula, Jr.
FOODSERVICE RESEARCH INTERNATIONAL, C.A. Sawyer
JOURNAL OF FOOD BIOCHEMISTRY, N.F. Haard and B.K. Simpson
JOURNAL OF FOOD PROCESS ENGINEERING, D.R. Heldman and R.P. Singh
JOURNAL OF FOOD PROCESSING AND PRESERVATION, B.G. Swanson
JOURNAL OF FOOD QUALITY, J.J. Powers
JOURNAL OF FOOD SAFETY, T.J. Montville and D.G. Hoover
JOURNAL OF TEXTURE STUDIES, M.C. Bourne, T. van Vliet and V.N.M. Rao

Newsletter

FOOD, NUTRACEUTICALS AND NUTRITION, P.A. Lachance and M.C. Fisher

OPEN DATING OF FOODS

by

THEODORE P. LABUZA, Ph.D.

Department of Food Science and Nutrition
University of Minnesota
St. Paul, Minnesota

and

LYNN M. SZYBIST

1701 N. Concord Road No. 9
Chattanooga, Tennessee

FOOD & NUTRITION PRESS, INC.
TRUMBULL, CONNECTICUT 06611 USA

FOOD & NUTRITION PRESS, INC.

6527 Main Street
Trumbull, Connecticut 06611 USA

Library of Congress Control Number: 2001135365

ISBN: 0-917678-53-2

PREFACE

In 1979 a government study was conducted to analyze various aspects of open-dating practices used by food companies and supermarket chains for commercial food products. One question addressed was why foods were not dated in the same manner as was used for pharmaceuticals and over-the-counter drugs and mandated by the Food and Drug Administration. At that time about 20 states had some form of dating requirements for refrigerated foods, many of which differed significantly from each other. While many of the findings of this 1979 study supported the introduction of mandatory open-dating legislation for foods at the federal level, the research committee concluded that the food industry lacked data to properly implement an efficient open-dating system at that time. However, they recommended the issue be addressed again in the future.

Since then, a significant amount of work has been done on developing shelf-life testing methods for foods which came from the foundation of work done on shelf-life testing of military and NASA space foods in Marc Karel's lab at MIT where I (Ted Labuza) was one of his students. This work inspired me to continue such research at the Univ. of Minnesota. During this time we also saw the introduction of simple devices that could be put on an individual food package, i.e., time-temperature integrators (TTIs) that actually showed the extent of degradation of foods that was dependent on time-temperature history. TTIs could provide the industry with sufficient means to generate truthful open dates on most food products. In this twenty-year period, significant mandatory dating requirements for foods appeared around the world, especially in the EU. With globalization of food distribution, this left the U.S. behind in providing a useful tool for the supermarkets and consumers.

The idea to reevaluate the open-dating practices of the U.S. industry was a direct result of the establishment by the Sloan Foundation of The Retail Food Industry Center (TRFIC now TFIC) at the University of Minnesota. TFIC is a Center that has the objectives to examine both the business practices of the retail food industry, as well as the food quality and safety issues resulting from the distribution chain. The Sloan Foundation has created over 13 different centers to study the practices of many different industries, e.g., banking, silicon chips, transportation, etc. When the Bil Mar incident of 20 deaths from consuming hot dogs with Listeria occurred, it was hypothesized by USDA that the food was consumed near or beyond the end of the shelf-life on the label and that temperature abuse may have occurred. Some manufacturers began to reduce their labeled shelf-life, although there was no real evidence that this was needed or would work. This then catalyzed TFIC to fund some studies to evaluate what has happened to open dating in the U.S. over the past 20 years and to determine what consumers feel about open dates on foods. This resulted in the research conducted by Lynn Szybist and created the foundation for this book.

Open dating plays a vital role in the distribution of food products from the farm or place of manufacture to the consumer's home. One principle established in the shelf-life research mentioned above, is that the true shelf-life is a function of the distribution conditions and can be looked at as the percentage of consumers a manufacturer is willing to displease. Thus one purpose of an open date is to give consumers enough time to purchase a food product and store it at home for a reasonable period of time before the product reaches the end of shelf-life in terms of some degree of quality change that is still acceptable.

If products are not stored or rotated properly in distribution, then older products or those of lesser quality and perhaps compromised safety because of abuse during distribution, will reach the consumers, indicating a serious problem in the distribution chain and causing loss of confidence in the food category. Presently, the lack of uniformity among manufacturers and across state borders has made the practice of open dating confusing and misleading for consumers, retailers, and the government. This book addresses these issues, and provides scientific and legal background to both evaluate and influence federally-regulated open-dating legislation in this country.

THEODORE P. LABUZA
LYNN M. SZYBIST

CONTENTS

CHAPTER 1

INTRODUCTION

Definitions

An *open date* or *open shelf-life date* refers to the practice of labeling a packaged food with a date that indicates when the product was packed, meant to be "sold by", or meant to be "used by" (Labuza and Szybist 1999a). *Shelf-life* is best represented as the end of consumer quality determined by the percentage of consumers that are displeased by the product (Labuza and Schmidl 1988). This definition accounts for the variation in consumer perception of quality and has an economic part, i.e., it is not possible to please all consumers; thus, one must establish a baseline of consumer dissatisfaction (Labuza and Szybist 1999a).

Purpose

A federally regulated open-dating system on food products, instead of the current somewhat random and non-uniform state-mandated system, would benefit today's consumers, retailers, and government agencies. Consumers have indicated a strong desire for open dates; it would enhance their ability to make educated choices about the freshness of the foods they consume. A mandatory/uniform system would also assist retail grocers with stock rotation, so that customers can be provided with the best products available. Finally, federal open-dating regulations across state borders would lessen burdens on interstate commerce. The potential benefits of this dating system outweigh the opposing points-of-view. This book analyzes data regarding the current open-dating practices and regulations of food products in this country, and proposes an open-dating regulation that would most benefit the American people.

Previous Research

In 1979 the Office of Technology Assessment (OTA), the research branch of the U.S. Congress committed to developing background papers, was charged with examining the effects and feasibility of mandatory open-dating legislation at the federal level. The committee chairman was Massachusetts' Senator Ted Kennedy, whose state's statutes at that time required the open dating of all food products. A task force was formed which included consumer representatives, food retailers, processors, wholesalers, scientific experts, and state and federal government officials. The committee analyzed the following areas concerning open dating: (1) consumers' perspectives on open-date labeling, (2) benefits and costs, (3) alternative systems and techniques, (4) alternative criteria and scientific tests to establish open dates, (5) enforcement mechanisms and liability related to open-date labeling, and (6) options available to Congress. OTA concluded that the industry lacked data to properly implement an efficient open-dating requirement at that time but recommended the issue be addressed again (OTA 1979). This book readdressed many of the same areas of concern from the OTA study.

Outline of Present Open Dating Situation

Chapter Two gives a brief history on the practice of open dating and discusses consumer demand for comprehensible dates. The problems with temperature abuse and food handling are discussed in Chapter Three along with the benefits associated with the implementation of time-temperature integrators on food products. Chapter Four explains how to properly implement an open date, and Chapter Five provides an extensive overview on current practices. Current regulations of both the U.S. and international practices are outlined and discussed in Chapter Six, while Chapter Seven presents a proposed regulation to mandate standardized open dating on all food products in this country. In Chapter Eight, past cases involving open dating in the judicial branch are presented while potential legal actions are described involving open dating and misbranded foods. The book concludes with the results from a consumer survey (Chapter Nine) focusing on the present status of consumer knowledge in regard to open dating and related issues followed by a brief summary (Chapter Ten).

CHAPTER 2

BACKGROUND AND RATIONALE

The History of Open Dating and Its Demand by Consumers

Open dating was established in the dairy industry in 1917 (Anon. 1979), and by the 1930s, *Consumer Reports* published consumers' desires for an open-dating regulation to indicate the freshness of their foods. Supermarket chains did not begin implementing some type of dating system, though, until the early 1970s (Seligsohn 1979). During 1979-80, although the Food and Drug Administration (FDA) had yet to propose any federally required open-dating regulations, a number of consumers, processors, and consumerist groups held hearings discussing its possible future implementation (IFT 1981).

Within the last three decades, extensive surveys and research on open dating have been more prevalent, and reports have indicated a high consumer demand for open-dating regulations. For instance, an A.C. Nielsen Co. (1973) report stated that many people looked on food packages for some type of date to aid them in selecting the freshest food. In another study, the Economic Research Service (ERS) of USDA and the Consumer Research Institute (CRI) conducted a consumer survey concerning food spoilage (USDA 1973). The results showed a lack of consumer confidence in the products they had purchased from the supermarket. While 93% of the respondents reported that they had not purchased any stale or spoiled products within the past year, many of them indicated a problem with the freshness of foods. Within the last two weeks prior to being surveyed, 18% of the customers purchased food which spoiled or staled before they expected. When a food was spoiled on the day that it was purchased, most consumers reported that they threw the product out rather than returning it to the store. This is in spite of the fact that 62% of the shoppers knew about the store money-back guarantee.

In the 1970s, as is also true now, many manufacturers preferred using code dates on their food packages. The purpose of a code date was to assist supermarket employees with stock rotation and as a means of lot identification in the case of product recalls. Although the everyday use of code dates was not directly intended for the consumers' benefit, a consumer group, the New York State Consumer Protection Board, published a book deciphering the meanings of the manufacturers' code dates. At that time, some supermarkets actually put several books in their stores so consumers could decipher the code dates on food products while they shopped. There were over 100,000 orders for this code book in its first year of publication (IFT 1981), although the book does not exist today.

The Food and Drug Administration (FDA) conducted its own survey in 1973 concerning open dating and published the results in the *Federal Register* (FDA 1979). This led to a federally sponsored project by the Office of Technology Assessment to determine if such dating needed federal regulation (OTA 1979). After surveying 1,374 grocery shoppers, FDA reported that 94% of those surveyed claimed to have noticed the open dating on some food products, and 75% of them used the date in making a purchase (Labuza 1982). In 1981, the Institute of Food Technologists' Expert Panel on Food Safety and Nutrition published a scientific status summary titled "Shelf-Life Dating of Foods" to inform professionals about open dating (IFT 1981).

The popularity of the open-dating issue seemed to lessen after 1980 until about 1996. At that latter time, *Supermarket News,* a publication targeted at the grocery industry, reported that based on a consumer study, peak freshness was the most important quality consumers looked for in food products (Dowdell 1996). In fact, fresh products in the meat department were the most important factor in determining where 73.6% of the surveyed consumers shopped (Stickel 1996). Also, when consumers saw a "sell-by date" on a food product, it heightened their confidence about the foods' freshness (Dowdell 1996). Although an open-date system does not guarantee the consumer that a food product is not spoiled, as would happen if it was improperly handled, a date can be used as an indication of freshness. Other research indicated that to increase effective communications with the consumer, simple and basic information works best. According to Joanne Gage (Vice President of Consumer and Marketing Services at Price Chopper Supermarkets, Schenectady, NY), the "sell-by" or "use-by" dates on pre-cut packaged items, like fresh-cut salads, are the most sought after information by the consumer (Williams 1998).

Efficient Consumer Response (ECR)

The Efficient Consumer Response (ECR) Working Group was developed in 1992 to form a "joint industry task force" among grocery retailers, distributors, suppliers and brokers to increase the competitive edge in the grocery industry. While looking at current practices in the industry, the overall goal of the group is to create potential opportunities in the grocery business and raise customer satisfaction without a huge financial burden.

While implementing ECR will mainly require reorganization of the internal and external structures of the grocery supply chain and changes in communication among those sectors involved, an efficient open-dating system fits right into the guidelines of this project. The following is Principle 1 of the "Guiding Principles of Efficient Consumer Response":

Constantly focus on providing better value to the grocery consumer: better product, better quality, better assortment, better in-stock service, better convenience with less cost throughout the total chain.

An efficient open-dating system would actually enhance the ECR system and increase its benefits to the store. The intangible benefits of ECR for the consumer, distributor and supplier are listed in Table 2.1. Many of the benefits of ECR also mirror the benefits of open dating (which is the focus of a later section in this chapter) (Kurt Salmon Associates 1993).

TABLE 2.1
INTANGIBLE BENEFITS OF ECR

Consumer	Increased choice and shopping convenience, reduced out-of-stock items, fresher product.
Distributor	Increased consumer loyalty, better consumer knowledge, improved supplier relationships.
Supplier	Reduced out-of-stocks, enhanced brand integrity, improved distributor relationships.

(Kurt Salmon Associates 1993).

Hazard Analysis and Critical Control Point (HACCP)

Beginning in 1959, the Hazard Analysis and Critical Control Point System was designed by Dr. Howard Bauman of the Pillsbury Company to create a system to guarantee the safety of food taken on-board manned space flights by NASA. About 25 years later, the food industry seriously started looking at the system for more general implementation. The concept of identifying and controlling the source of contamination by establishing CCPs (Critical Control Points) *before* the product reaches the consumer allows food manufacturers to take control measures to detect and prevent the production or introduction of unsafe food (Hartman 1997).

FDA and USDA have both published regulations mandating the establishment of HACCP for several distinct classes of perishable domestic and imported food processes. By January 25, 2000, all meat and poultry processing plants were required to employ a HACCP system (FSIS 1996). Regarding FDA, seafood is currently mandated to carry a HACCP plan with the mandatory implementation for orange juice manufacturers under consideration (FDA 1994; FDA 1998; FDA 1996). HACCP plans are also strongly encouraged for foodservice establishments (FDA 1995).

HACCP is also making its presence felt at the international level. In Australia, for example, the country is considering the adoption of HACCP systems to incorporate food safety plans into the national food hygiene standard. Such action is warranted based on a study published about food temperature abuse in five Australian facilities (Krassie 1999). It was found that poor temperature control was a major factor contributing to the rise in foodborne illnesses in the country. According to the Australian Food (General) Regulations Section 80 (6)(a) and (b), the minimum temperature standard for hot foods is 60°C (140°F), while the standard for cold food is a maximum of 5°C (41°F). At three of the four testing sites, however, all of the cold food exceeded the maximum standard at the point of delivery. At two facilities monitored for hot vegetable temperatures, 85% of the samples at point of delivery did not meet the temperature standard.

Many other countries, such as those in the European Union (EU), New Zealand, Canada, Japan, Egypt, and South Africa are also implementing or considering the HACCP system into their own food safety regulations (Bernard 1997). Consequently, HACCP-oriented training courses and guidelines are being sponsored by trade and professional associations worldwide, and the National Sanitation Foundation is integrating ISO-9000 principles with HACCP to initiate HACCP-9000, a food safety, third-party auditor (NSFI 1997).

While HACCP is currently being implemented to monitor CCPs and to ensure a safe food during processing, handling, and distribution of the product before it reaches the consumer, steps must be taken to reduce food mishandling once the product reaches the consumer's hands. During the last several decades, the education of children and the domestic aspects of the typical American family lifestyle have changed dramatically. For instance, students are studying less home economics-type classes at school, and most mothers have joined the work force. As a result, fast food and frozen dinners are becoming more prevalent in American homes, and many of today's children are growing up with a lack of food handling knowledge in the kitchen. Therefore, a basic HACCP system must be extended to view today's kitchen as a critical factor in the food chain.

Potential Benefits of Open Dating

As with every issue, there are pros and cons associated with open dating. The following "pros" reflect the advantages of a mandatory, uniform regulation at the federal level. Some of the potential benefits include the overall increased awareness in food safety, food quality, nutrition, functional properties, stock control, education, and updated regulations.

Food Safety. A regulated open-dating system serves as a guide for "first in-first out" practices at the retail level. As stated in Grocery Manufacturers of America (GMA) v. Department of Public Health (393 NE 2d, 881, 1979), the dating of foods helps deter the possible consumption of old products which may have a higher risk for non-safe agents (microorganisms).

In 1999, the Center for Disease Control (CDC) reported 21 deaths (15 adults and 6 miscarriages/stillbirths) across the nation from an outbreak of *Listeria monocytogenes* (CDC 1999). The source of the foodborne illness was ready-to-eat meat products, such as hot dogs and luncheon meats, consumed very close to the stamped "use-by" dates. If the products were temperature-abused, however, conditions may have allowed the pathogenic organisms to achieve an infectious level before the end of other food quality attributes, as was inferred by Thomas Billy, head of USDAs Food Safety and Inspection Service (Silverman 1999). To avoid such incidents in the future, federal rules governing how the "sell-by" and "pull-by" dates are determined on packaged foods is necessary (Anon. 1999a). A shorter shelf-life on such products will increase the products' turnover rates and lessen the chances of microbial infections.

Overall Quality. Under the proper conditions, an open date could indicate the end of acceptable food quality and would increase consumers' confidence in the food products they purchase. Out-dated foods are more likely to have lower sensory qualities relating to color, texture, odor and flavor changes (OTA 1979).

An in-store experiment conducted by USDA supported the concept of increased consumer confidence in food quality when open dating was introduced. In foods commonly cited for spoilage or staleness, such incident complaints at the experimental store were reduced by 50 percent after open dating was implemented. The store also experienced decreased financial losses and package rehandling. The spoilage complaints decreased in both the open-dated and non-open-dated products. Although the open-dating system did not reduce food spoilage, it may have attributed to the consumers' increased confidence in the overall freshness and food quality of the food products sold at that store (OTA 1979).

Nutrient Levels. Based on the temperature and humidity conditions during distribution, degradation of several essential amino acids and vitamins (such as A, B, and C) may cause some food products to have nutrient levels below the level stated on the label (IFT 1981). Under normal conditions, the open date can be set as the time when the nutrient level goes below the legal level stated on the label for the most labile nutrient, e.g., vitamin C in refrigerated orange juice. It is not clear, however, how companies base their open dates in compliance with nutritional labeling laws as defined in 21 CFR 101.9 (g).

Functional Properties. During storage, some foods will lose their functional properties (Anon. 1971). Baking yeast, for example, will lose some of its leavening capabilities; eggs will decrease in their whipping abilities; and pre-emulsified foods, such as salad dressing, may separate over time, especially if temperature-abused. With respect to eggs, it is suggested that they will fall below the labeled grade (A, B, etc.) within 30 days at refrigerated temperatures.

Retail Stock Control. With more than 100,000 different code dates used by the food industry, it is unreasonable to expect food distributors and supermarket employees to decipher every one of them. In 1972, Keith Ford of the Minnesota Office of Consumer Services found that in a survey across the State of Minnesota, 44% of the infant formula on the grocers' shelves were past the "use-by" date. An astounding 64% of the store managers could not even read the date codes on the food packages; therefore, they were not rotating the stock to allow for selling of the oldest product (Labuza 1982). In addition, of the 25 stores surveyed, 100% had out-of-date products on the shelves.

Current dating practices continue to be confusing to retailers and consumers as shown in Table 2.2 with respect to yogurt (Labuza and Szybist 1999a). The table shows various dating systems among yogurt manufacturers and differences within the same brand in regard to the suggested storage temperature. More importantly, one label recommends proper storage at 44°F, which is above the safe and recommended refrigerator temperature of 41°F. There was no explanation of the date on containers of Dannon yogurt. To understand the date, one must dial the toll-free number on the container, which is inconvenient for both consumers and stock rotators at retail level.

A recent policy change by Procter & Gamble (P&G) Co. for returned goods may increase retailers' need for understanding code dates. The company will no longer be responsible for goods that the retailers cannot sell, including those items with limited shelf-life. Instead, P&G will pay initial lump sums to retailers on a quarterly basis to cover the cost of old or damaged goods. As Narisetti (1997) wrote, "...it's likely that P&Gs payment plan will seem like a bonus to efficient retailers and prove costly for those with poor ordering or handling systems." Other companies are expected to follow P&Gs lead, which would increase the need for a legible open date printed on prepackaged food in order to improve inventory control (IFT 1981).

Home Stock Control. Readable dates not only assist supermarket personnel in stock rotation, but also lessen the chance of consumers purchasing food of substandard quality (Anon. 1979). The problems of careless storage practices at home could lead to food products of lower quality and could potentially lead to a food safety problem.

TABLE 2.2
CURRENT OPEN DATING PRACTICES ON YOGURT CONTAINERS
(Collected on 1/13/98)

YOGURT		
Product & Address	***Open Date***	***Printed Dates***
(5 oz) **Kemps Nonfat** (Various Flavors) **Yogurt** Marigold Foods, Inc./ General Offices Mpls., MN 55414	Remains wholesome one week after date on carton.	JAN. 28 to FEB. 17
(32 oz) **Kemps Nonfat Vanilla Yogurt** Marigold Foods, Inc./ General Offices Mpls., MN 55414	When properly refrigerated between 33 & 44°F, this product will retain its wholesomeness for 1 week beyond date on carton.	FEB. 5
(8 oz) **Gaymont Lowfat** (Various Flavors) **Yogurt** (Old Home Foods) Old Home Foods St. Paul, MN 55103	Quality assured 7 days beyond code date on side if properly refrigerated (40-44).	FEB. 5 to FEB. 19
(24 oz) **Gaymont Lowfat** (Various Flavors) **Yogurt** Old Home Foods St. Paul, MN 55103	Quality assured beyond code date on side if properly refrigerated (35-40).	FEB. 5 to FEB. 12
(8 oz) **Dannon Yogurt** Dannon Company, Inc. Tarrytown, NY 10591	---	JAN. 17 to FEB. 8

(Labuza and Szybist 1999a).

A survey by Beard III was conducted in 30 households to evaluate the home storage practices of the participants (Beard III 1991). The results from Beard's study (which are further discussed in Chapter Nine) indicated that the home storage rotation practices of the participants were poor. Since many households do not date their food purchases, an open date would be useful in managing household stocks (Beard III 1991).

Educating the Consumer. A 1979 FDA study found that only 1.3% of the respondents felt confused about what the open date on food packages represented. After further questioning, however, it was revealed that most of those surveyed did not actually understand the meaning of the dates (IFT 1981). A 1979 OTA study also gathered information concerning consumer understanding of open dates as shown in Table 2.3. The participants were mixed as to what they understood the date to mean on milk, breakfast cereal, and ground beef packages.

TABLE 2.3
CONSUMER UNDERSTANDING OF FRESHNESS DATES
(Percentage of Respondents)

	MILK	*BREAKFAST CEREAL*	*GROUND BEEF*
When it was packaged...	9	8	***34****
Last day it should be sold...	***74****	35	31
Last day it should be used or eaten...	15	***26****	9
Have never noticed a date on a package of this product...	2	31	26

Note: Percentages in boldface with asterisks indicate correct answers (OTA 1979).

About three-quarters of the shoppers correctly identified the date on milk; only one-quarter of the answers for breakfast cereal and one-third of the answers for ground beef were accurate. A 1992 Minnesota/South Dakota Dairy Center study showed that 94% of those surveyed thought that the open date was extremely important. But 25% of that population doubted the reliability of the date, while 61% admitted that they did not fully understand it (Sherlock and Labuza 1992).

In an effort to educate its readers, *National Enquirer* published a short article explaining the different meanings of commonly used open dates (Anon. 1999b). Such media efforts are important, especially after several startling studies demonstrated the need for better consumer education. In a 1997 joint survey by *Prevention* magazine and *NBC Today*, for example, results suggested that most people thought the dates were related to either the last date to safely sell the product or the last date to safely eat it (Anon. 1997a). Open dates, however, are not a guarantee of food safety. With consumers understanding that an open date is only an indication of freshness and that the date written on the product is not necessarily the end of product quality, this may lead to a reduction in food waste and perhaps less outbreaks of foodborne illness.

Reducing Perishable Food Waste. Recent evidence suggests that 50% of the consumer dollar is spent on refrigerated perishable food products (Anon. 1998a). Along with this, there is also a substantial amount of food waste generated. In 1995, for example, fluid milk accounted for an estimated 17.4 billion pounds or 18.1% of the total edible food lost by retailers, foodservice and consumers in the U.S. (Kantor *et al.* 1997). In the present study, the open dates of milk containers were observed on November 25, 1997, at a local supermarket outside of St. Paul, MN. The "sell-by" dates on the milk containers ranged from November 28 to December 12, indicating that the products should be sold within three to 17 days to still maintain a reasonable time before reaching the end of shelf-life at home (Labuza and Szybist 1999b). Dating of milk, however, is not mandatory in every state. A consistent and consumer-friendly open-dating system is necessary to assist with efficient stock rotation and to give Americans an indication of whether their purchased milk and other food products have an estimated shelf-life of several days or a couple of weeks. A mandatory dating system would also increase understanding of open dates by the retailers and consumers, i.e., that the date on milk represents the recommended day of selling the product off of the shelves and not the end-of-shelf-life.

Decrease Overall Misconceptions. In an Institute of Food Technologists (IFT) press release to newspaper editors (December 15, 1998), some "food safety" tips were given, which included the following statement: "**Don't debate the date.** Don't buy food past its expiration or sell-by date, or food that will not be used by its sell-by date" (IFT 1998). The tip implies that all products consumed before their dates are safe, which is untrue if the food is temperature-abused in transportation or storage. It also implies that the food becomes unsafe at the end of the "sell-by" or "use-by" date. Manufacturers generally base their dates on quality not safety, especially with respect to the "sell-by" date, which was intended to help retailers in stock rotation. Therefore, this recommendation, while useful, is not totally truthful. Only by knowledge of the time-temperature history during distribution would one be able to make some judgement on safety.

Reducing 'Out-Dated' Legislation. The open-dating legislation is outdated in several states. In Minnesota, for example, perishable foods are required to have an open date, but fresh fruits and vegetables, meat, and poultry are excluded. However, with advances in technology, certain foods, including fresh fruit and vegetables, meat, and poultry can be packaged under controlled atmospheric/modified atmospheric (CAP/MAP) conditions. The process increases the shelf-life of the food, but also adds the possible presence of anaerobic pathogens if the food package is mishandled. Because of the short experience of these foods among retailers and consumers, ensuring proper rotation is necessary to help move the product from production to table more rapidly. An open date must be mandated for CAP/MAP foods regardless of the food content.

Other out-dated legislation includes the open-dating laws of Iowa and Nebraska, which require the dates on reduced oxygen packaged foods (such as the CAP/MAP packaged foods) not to exceed 14 days. Adhering to these laws would make some refrigerated filled pasta products illegal in those states if held for sale beyond this time even though they have about three months shelf-life under normal refrigerated conditions.

Potential Disadvantages of Open Dating

Not everyone is in favor of a mandatory open-dating system. About 41% of the states (including the District of Columbia) still do not require an open date on any food products, and many small companies and food producers are against the "inconvenience" of open dating. More serious drawbacks of an open-dating system include the premise that it is a guarantee of food safety protection as well as a greater expense for consumers.

Lack of Temperature Control. Concerning food safety, an open date cannot prevent failures during or after processing or improper practices in the distribution chain, such as temperature abuse, which is further discussed in Chapter Three (IFT 1981). As stated in a University of Minnesota study on food shelf-life, "...if the food is temperature abused, an open date is meaningless and in fact is a false sense of security" (Taoukis *et al.* 1991). In other words, an open-dating system would provide little to no help in indicating that levels of microbial pathogens were present in abused food products (IFT 1981). The date itself may be used as justification to sue when a food poisoning outbreak occurs irrespective of the actual cause of the outbreak.

Financial Influences. The financial aspects of implementing open dates are a necessary consideration. In 1979, experts at the request of the Office of Technology Assessment (OTA) were consulted to discuss the costs of open dating. The issues that were investigated included the costs of establishing the shelf-life of the food, the cost of putting the date on the package, and any potential enforcement costs.

Shelf-life Tests. Shelf-life determinations were estimated to cost about $100,000 per item for perishable foods. This included costs for an investigator, a technician and a one year facility charge. Non-perishable foods would have cost approximately $200,000 per food item because shelf-life testing would require about two years, and there are many more variables to consider, e.g., see Labuza and Schmidl 1988. Although these costs would be one-time costs, shelf-life adjustments would be necessary with changes in product formulation, packaging, and mode of distribution.

Printing Equipment. Dating a package would most likely require additional printing equipment, especially if not combined with the printing of the code date which is generally not in an easy-to-see location on the package. The type and design of the food package and location of date would affect costs. The estimated costs for open dating on cans was about $1,000-$3,600 per machine. Equipment for perishable foods varied in costs between $1,500-$15,000 per food product (OTA 1979). However, with present laser printer technology, these costs might be less today.

Enforcement. Enforcement costs would be difficult to estimate without knowing the enforcement system in need of being implemented. One option would be a self-enforcing system, where there are no penalties for out-of-date products. Instead, customers would enforce the open-dating system by refusing to pay full price for such items. Obviously, however, if the product was temperature abused, this could lead to food safety problems. If legal penalties were to be enforced, the costs were estimated to be over $500,000 per year and perhaps much more (OTA 1979). Note that all of these financial figures were calculated around 1979, so some of the costs could be considerably higher today.

Consumer Sorting Habits at the Retail Level. From the supermarket standpoint, the cost of stock rotation might be considerable to prevent people from sorting for the youngest date and increasing food waste. Dairy and bakery products are two of the largest contributors to food loss, because products which are still safe to eat and of acceptable quality are often removed from supermarket shelves once reaching the "sell-by" date. In a 1971 study, 62% of the 628 people in the survey stated that they sometimes sort through packages to find the freshest product. From that same group, 74% claimed that while sorting through dated products, they would usually find some products that were fresher than others (Anon. 1971).

CHAPTER 3

TEMPERATURE ABUSE AND TIME-TEMPERATURE INTEGRATORS

Temperature Abuse

As stated by Dr. Ted Labuza, "Shelf-life is not a function of time alone, rather it is a function of the environmental conditions and the amount of quality change that can be allowed" (Labuza 1982). These "environmental conditions" relate to the temperature of food products during storage and distribution.

The maximum temperature recommended for chilled foods in warehouses, trucks, and retail displays by the National Food Processors Association (NFPA) is 40°F (Brody 1997). NFPA has also published information for the optimum temperature ranges of various food products during distribution and storage as shown in Table 3.1. In the revised Food Code (1997), FDA established 41°F as the recommended maximum temperature for retail establishments handling meat, fish, poultry, delicatessen products and pre-cut produce (Brody 1997).

Maintaining the optimum temperature for food storage and distribution can be costly. Besides the possible expense of employing new technology to maintain proper temperatures, retailers and distributors face the financial responsibility of a 10% increase in total energy costs for every 5°F the refrigeration temperature is lowered (FNQUEB 1999). The economic costs should not overstep the boundaries of safety, but studies show that the safety issue is not being taken seriously and proper steps in temperature control are not being enforced.

TABLE 3.1
OPTIMUM TEMPERATURES FOR CHILLED FOODS

Food Product	*Optimum Temperature Range*
Dairy Products	32-40°F
Meat & Poultry	30-34°F
Seafood	30-34°F
Salads	32-40°F

(Brody 1997).

Examples of Temperature Abuse.

1980s Tropicana Products, Inc. In the 1980s, the Tropicana Products, Inc. discovered that 72% of consumer complaints on chilled juices were connected with temperature abuse (Kalish 1991). When examining this problem, it was found that the temperatures during distribution of the juice reached 45°F in some cases. The recommended temperature range for proper storage of the juice is 32-38°F to maintain high quality. During storage, the average temperature of the juice was about 44°F and went as high as 56°F. Within retail stores, only 37% of the products were stored at the proper temperature. Rotation practices in the display cabinets were also found to be poor.

1989 Audits International Survey. Temperature measurements were recorded of 1,000 refrigerated food items at three points: at the retail level, when the products reached the consumers' homes, and after 24 hours. Delicatessens in the study had refrigerated foods stored at temperatures ranging from 34 to 71°F; the average was 47°F. The average home refrigeration was 43°F (Audits International 1990). Both of these averages are above the recommended 40-41°F temperatures for refrigerated food, meat, poultry, and eggs.

1991 Tropicana Products, Inc. In 1991, the Tropicana Products, Inc. conducted a time-temperature experiment measuring flavor quality. Juice was rated on a scale of 1-9 with scores below 5 being unacceptable. The results showed dramatic differences in quality with temperatures only 10°F different. These changes of the shelf-life as a function of time-temperature are a function of its temperature sensitivity factor (Q_{10}). Juice held at 45°F fell to a 5.1 score after only 49 days, while juice held at 35°F, i.e., 10°F lower, remained above 6.0 for the entire 63-day shelf-life (Kalish 1991).

1992 Vaccines and Temperature Abuse. Lack of temperature control is not just a problem in the food distribution chain. Bishai *et al.* (1992) found that of 50 pediatric clinics in the Los Angeles area, only 16% of vaccine (measles, mumps, rubella) storage coordinators could cite appropriate storage temperatures for vaccines. Eighteen percent of coordinators were unaware that temperature abuse would destroy the effectiveness of the vaccines. Refrigerator thermometers were checked once weekly in only 20% of the offices, 22% of the refrigerators were above the required temperature range, and 16% of the offices stored vaccine unrefrigerated.

1997 The Refrigeration Research and Education Foundation (RREF). An RREF study in 1997 reported that approximately 20% of retail chilled display cabinets operated at temperatures above 50°F and that over two-thirds of the chilled food retailers did not monitor the expiration dates on their products (Brody 1997). Such studies illustrate a potential for growth of pathogens that might cause food poisoning especially in abused chilled food products that are minimally processed. At temperatures above 40°F, pathogenic anaerobic microorganisms are capable of growth and toxin production. Between the temperatures of 40 and 55°F, nonproteolytic anaerobic microorganisms may grow. Such spoilage is not detectable by virtue of smell (Brody 1998).

1998 Audits International Study. At the International Fresh Cut Produce Conference in San Diego, Dr. Daniels of Audits International reported a follow-up study in regard to temperature measurements and refrigerated food storage (Anon. 1998b). In a sample of 98 supermarkets, the mean temperature in the refrigerated deli case was 46°F with 10% of them at a mean of 58°F. In the produce section, packaged salads were also at a mean of 46°F. In both cases, unacceptable temperatures would cause more rapid deterioration of prepackaged foods and could lead to a safety issue.

1998 The University of Nottingham, England. Led by Dr. Angela Johnson, this study concentrated on the elderly, who represent the highest risk of mortality from food poisoning (Johnson *et al.* 1998). The participants' food safety knowledge and practices were evaluated. While the results for overall understanding of the "sell-by" and "use-by" dates were acceptable, 45% of them could not read the dates because of poor eyesight and because of small and hard-to-read print. Of even greater concern, however, was the fact that 70% of the participants' refrigerators were too warm to safely store food.

Transportation and Other Concerns. In 1995, concerns regarding lack of proper refrigeration led to the Transportation Analysis Group (TAG). TAG was formed by the U.S. Department of Transportation (DOT) and the Food Safety and Inspection Service (FSIS) of USDA to discuss potential solutions based on the HACCP concept. After an open meeting in Washington D.C. in June of 1996, FDA announced an advanced notice of proposed rulemaking jointly with FSIS and DOT (FDA 1996). In this document, they note that "...post processing transporters, storage operators and retail stores, restaurants and other food service sectors are important links in the chain of responsibility for food safety." Because of this and the noted lack of temperature control, these regulatory agencies proposed six possible steps that could be taken. Note that this includes holding at the grocery store level:

(1) Setting mandatory temperature performance standards, e.g., a maximum of 41°F or 45°F for potentially hazardous foods, (2) Requirement of shipper record keeping, (3) A mandatory HACCP-type system, (4) Voluntary guidelines

such as in the Association of Food and Drug Officials (AFDO) "Guidelines for Transportation of Food", (5) A combination of approaches, and (6) No federal initiative.

All of these alternatives would have an impact on open dating, but as of present, there has been no proposed regulation. Note that in 1990, a comment by Dr. Ted Labuza of the University of Minnesota in regard to a USDA/FSIS advanced proposed rulemaking on refrigerated foods suggested the use of time-temperature integrator (TTI) tags (Anon. 1990). TTIs are one method that could potentially solve part of the problem of temperature abuse for chilled foods and would fall under the HACCP approach.

An open date will not protect consumers from microbial threats, but it can be useful as a guide. If the product is kept at ideal temperatures and conditions throughout its lifespan, then the food will most likely be safe. In 1998 an article in *Newsweek* reported that to guarantee temperature abuse has not occurred, "tell the truth" tape or time-temperature integrators (TTIs) are being designed to signal a premature end of shelf-life (Taoukis *et al.* 1991; Springen 1998). After consumers check the TTI, they will have confidence that their product has not been temperature abused, and the open date will indicate approximately how long the product will remain fresh under proper conditions. Such information is useful at both the retail and consumer levels, since many consumers are not aware of the actual storage conditions in their own homes and local supermarkets.

Time-Temperature Integrators (TTIs)

Definition. Time-temperature integrators (TTIs) are small, physical devices that are placed on food packages to measure the temperature history of a product and indicate a definitive change at the end of shelf-life (Taoukis *et al.* 1991). Through "integration" of the time-temperature exposure, TTIs are reliable indicators of end-of-shelf-life for food products if they have similar temperature sensitivities (Q_{10}) as for the deterioration mechanism.

The three major manufacturers of TTIs are 3M, LifeLines, and VITSAB as shown in Fig. 3.1.

Potential for Use. Loss of quality and "freshness" are not solely a function of time, but depend on the control of temperature, humidity and light during distribution. Both the kinetics of deterioration and the product's history of exposure, including temperature abuse, must be incorporated into shelf-life dating (OTA 1979). Due to such obstacles, the OTA Committee in 1979 did not pursue a federally-mandated open-dating legislation, but suggested that research be done on TTIs.

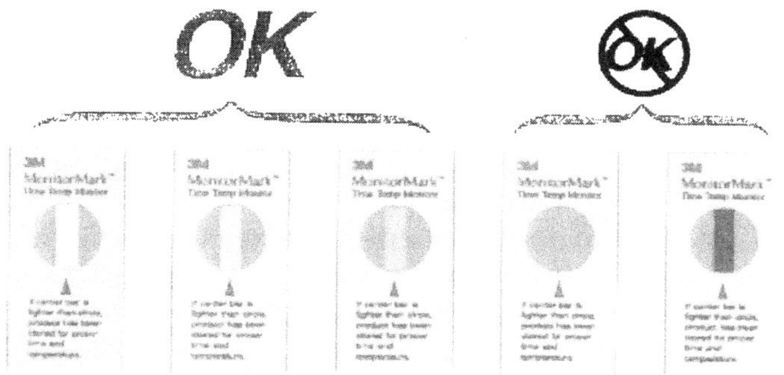

a. 3M Monitor Mark showing change in color as the product ages

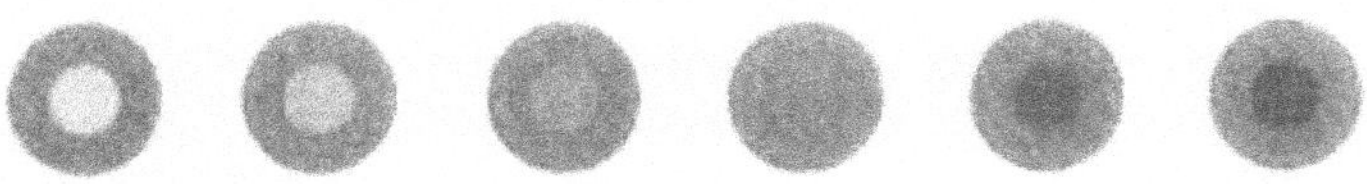

b. LifeLines TTI showing change in color as the product ages

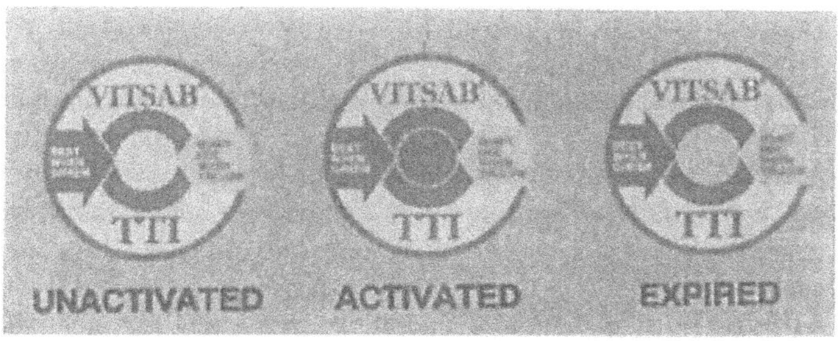

c. VITSAB TTI showing change in color as the product ages

FIG. 3.1. THREE TYPES OF TTI

With the proper implementation of an open date along with a TTI, the degradation of a food product is monitored as a function of both time and temperature (Springen 1998). TTIs can increase the effectiveness of quality control in distribution and during storage (Taoukis *et al.* 1991) because of their ability to indicate shortened shelf-life due to temperature abuse. The open date gives the consumer a sense of the shelf-life left for a product. But if the TTI indicates the end-of-shelf-life much earlier than the open date, there may be a distribution or storage problem of which the manufacturer, retailer and consumer may not have been aware.

The devices can be used on individual consumer packages, so they establish a control system because not all products will receive uniform handling, distribution and time-temperature effects (Taoukis *et al.* 1991). Therefore, TTIs can increase the effectiveness of quality control in distribution, stock rotation practices of perishable foods in grocery stores, and efficiency in measuring freshness by the consumer (Sherlock *et al.* 1991).

Internationally several food manufacturers are beginning to implement TTIs on their food packages. In September of 1999 Italy made the first major attempt towards mandatory usage of such devices. The Italian Government will decide the fate of a proposed regulation entitled "New sanitary regulations aiming to assure a right and safe preservation of pre-packaged food products with a short shelf-life" (Appendix B). In the event that the proposal becomes law, the Italian Ministry of Health along with the Ministry of Industry would define standards for TTIs, and the devices would be calibrated according to the shelf-life of each food product regulated under this proposal. The legislation would also require standardizing features and performance requirements for the device and mandating suitable settings for placing the TTI on the food package. With a durability date (open date) required by the EU and TTIs on perishable food items, Italian shoppers may soon have the advantage of choosing fresher and higher quality foods compared to any other industrial society in the world.

TTI Studies. Taoukis and Labuza (1989a, b) found that for the most part, the commercially available TTIs are both reliable and applicable for use in combination with open dating of refrigerated foods. Malcata (1990) reported that although the tags respond more quickly to temperature abuse than the actual food, the response is on the conservative side of safety, i.e., the tag color indicates end of shelf-life before the food is actually spoiled. The Campden Food and Drink Association in the United Kingdom has developed technical standards for the evaluation of TTIs (Campden 1992).

According to LifeLines, their experience has indicated a decrease in product returns with the implementation of TTIs because the visual monitor for temperature abuse increases accountability in handling the product along the distribution and storage chain (Anon. 1999c). Cherng and Zall (1989) have also determined the LifeLines system to be an inexpensive method to monitor the performance of refrigerated cabinets during commercial distribution and in vending outlets. Regarding the VITSAB TTI, Leak and Ronnow (1999) concluded that the TTI could be used as a CCP (Critical Control Point) in receiving ground beef products and beneficial in monitoring inventory and overall quality management in meats.

In the dairy industry, the 3M and i-point products were found to be very effective and made it "possible to replace the sell-by date on market milk" (Mistry and Kosikowski 1983). Another study concerning the effectiveness of TTIs and dairy products concluded that "Although the tags did not sufficiently predict the sensory endpoint, it is still thought that tags would be beneficial to the dairy industry when used along with a 'Use-by date'" (Duyvesteyn 1997). Duyvesteyn focused on the reliability of the TTIs, but its use in conjunction with a "use-by" date will not be effective if there are no standards for determining a proper date.

In 1988, Find/SVP, a marketing consulting firm, concluded that the use of TTIs had great potential for use in the food business (Find/SVP 1988). A survey by the Business Marketing Research Inc. found that consumers would prefer clear, consumer-readable indicators of time and temperature to measure the freshness and safety of their perishable food products (Sherlock and Labuza 1992). In a consumer survey published by LifeLines, consumers indicated a preference to seeing both the open date and the TTI. The TTI would eliminate the need for shorter open shelf-life dates, as practiced by many manufacturers in anticipation of temperature abuse. This would also give retailers a longer period of time to display and sell their products, which would reduce food waste (Anon. 1999c).

Food Safety and Reduced Oxygen Packaging. TTIs are recommended to monitor the potential danger of temperature abuse for controlled atmosphere/modified atmosphere packaged (CAP/MAP) prepared meals. Improper conditions can lead to the growth of harmful pathogens, such as *Clostridium botulinum, Listeria monocytogenes,* and *Salmonella enteritidis*. In one reported case, TTIs may have prevented four people from becoming ill from *C. botulinum* (a potentially lethal pathogen) after consuming a temperature abused CAP/MAP shredded cabbage product (Sherlock and Labuza 1992). A study by FDA for potential outgrowth of *Clostridium botulinum* and toxin production has also suggested the use of time-temperature integrator tags as a way to monitor the storage abuse on individual packages of vacuum packaged fish (Skinner and Larkin 1998).

Current Uses. Due to strong private label product lines in Europe, TTIs are more prominent with European food retailers, who focus on creating a point of difference. During the past few years, the American supermarket competition for market share has been heavily based on pricing. Perhaps as more American chains develop strong private label programs as part of their business strategy, TTIs may become a more integral part of the U.S. food chain (Anon. 1999c).

The use of TTIs on meats, however, has been encouraged in the U.S. FSIS has recommended monitoring the temperature of meats in the processing room during the entire grinding process as established in the "Guidance for Beef Grinders to Better Protect Public Health" (Guidance for Minimizing Impact Associated with a Food Safety Hazard in Raw Ground Meat and Other FSIS Regulated Products). The document specifically mentions the use of TTIs on packages as an indicator of adequate temperatures of the meat during storage, distribution, and display of the products in grocery and other retail establishments (FSIS 1998a).

The military is presently looking into the use of VITSAB tags. The U.S. military spends about $1.7 billion annually to feed its active duty troops. An efficient indicator of microbial growth in temperature-sensitive and perishable foods is necessary to feed the troops in various conditions, such as desert weather to arctic conditions (Cox 1998). According to LifeLines, the U.S. Army already requires all of its co-packers of Meals-Ready-(for)-Eating (MREs) to implement the *Fresh-Check* TTI on every carton of product (Anon. 1999c). The LifeLines TTI is also being used on several perishable products by foodservice companies, such as Marriott Corp. and Outback Steakhouses. Supermarket chains, such as Monoprix (France), Trader Joe's (U.S.) and Continente (Spain), now require LifeLines TTIs on selected refrigerated, private-label products (Anon. 1999c). Eatsy's in Dallas, Texas, was also scheduled to begin using LifeLines' tags on their deli and refrigerated items sometime in 1999.

The 3M TTI was test marketed in many Cub Foods Stores in the Minneapolis/St. Paul, Minnesota area during November of 1998 on prepackaged CAP/MAP hamburger and generated much interest. Unfortunately, due to other marketing practice changes, it was not possible to determine directly the impact of the tag on consumer behavior. A dairy products manufacturer in the Twin Cities area has also expressed interest in the use of TTIs on specialty milk products.

CHAPTER 4

ESTABLISHING AN OPEN DATE

Modes of Deterioration

Potential forms of food deterioration are an important factor to consider when establishing an open date and managing the distribution chain (including holding practices at the supermarket level). A food begins to deteriorate as soon as it is packaged, but the rate of deterioration depends on a number of factors (OTA 1979). In 1979, OTA developed a table summarizing the potential types of deterioration for specific food products. As shown in Table 4.1, a food's primary mode of deterioration depends on a variety of characteristics: food composition, chemical constituents, enzymatic activity, processing technique, packaging use, and distribution conditions. In some cases the process of deterioration may be held up by proper handling techniques by grocers and consumers or by manipulations by the food industry.

Senescence. Senescence refers to the natural enzymatic activity of aging of a harvested or slaughtered product that utilizes carbohydrate and nutrient stores within the food. This process is a major influence on determining the open date for fruits and vegetables, whole grain cereals, meat, poultry, fish and sometimes even dairy products. Concerning fruit, it is a beneficial process when the product must be transported over significant distances, since these products are picked before peak quality. During distribution, senescence allows the fruit to age, as well as repair itself from postharvest damage and to reach maturity; whereas if the fruit was picked at optimum ripeness, the product would rot before reaching the marketplace. Senescence is also involved in the aging process, which allows meat to reach a desirable level of tenderness (Labuza 1982).

The fate of these raw foods, however, is the enzymatic biochemical process leading to degradation (loss of color, flavor, texture, and nutrients). The breakdown leads to tissue damage and increases the products' susceptability to microorganisms and quicker decay.

The following are common methods used to decrease the rate of degradation due to senescence:

- Decrease temperature (Labuza 1982)
- Use of controlled atmospheric/modified atmospheric packaging (CAP/MAP) technologies
- Use of edible barriers
- Control of ethylene (a plant growth stimulant-hormone)

TABLE 4.1
MAJOR MODES OF DETERIORATION, CRITICAL ENVIRONMENTAL FACTORS, SHELF-LIFE, AND TYPE OF OPEN DATING BY FOOD PRODUCT

Food Product	Mode of Deterioration *(assuming an intact package)*	Critical Environmental Factors	Shelf-life *(average)*	Date Most Suitable for Product	Additional Information
		Perishable Foods			
Fluid Milk and Products	Bacterial Growth, Oxidized Flavor, Hydrolytic Rancidity	Oxygen, Temperature	7-14 Days at Refrigerated Temperature	Sell-by	Length of Time Product Can be Stored at Home
Fresh Bakery Products	Staling, Microbial Growth, Moisture Loss Causing Hardening, Oxidative Rancidity	Oxygen, Temperature, Moisture	2 Days (Bread)/ 7 Days (Cake)	Sell-by	---
Fresh Red Meat	Bacterial Activity, Oxidation	Oxygen, Temperature, Light	3-4 Days at Refrigerated Temperature	Pack or Sell-by[1]	---
Fresh Poultry	Pathogen Growth, ***Microbial Decay***	Oxygen, Temperature, Light	2-7 Days at Refrigerated Temperature	Sell-by[1]	Length of Time Product Can be Stored in Home Either Frozen or Refrigerated
Fresh Fish	Bacterial Growth	Temperature	14 Days When Stored on Ice *(marine fish)*	Pack (Catch Date)[1]	---
Fresh Fruits and Vegetables	Microbial Decay, Nutrient Loss, Wilting, Bruising	Temperature, Light, Oxygen, Relative Humidity, Soil & Water, Physical Handling	Depends on the Specific Commodity[2]	Pack[1]	---
		Semiperishable and Perishable Foods			
Fried Snack Foods	Rancidity, Loss of Crispness	Oxygen, Light, Temperature, Moisture	4-6 Weeks	Sell-by or Best-if-Used-by	Home Storage Information such as "Store in a Cool, Dry Place"
Cheese	Rancidity, Browning, Lactose Crystallization	Temperature	Processed Cheese 4-24 Months/Natural Cheese 4-12 Months	Best-if-Used-by	---
Ice Cream	***Graininess Caused by Lactose Crystallization,*** Loss of Solubilization *(caking)*, Lysine Loss	Fluctuating Temperature *(below freezing)*	1-4 Months	Sell-by or Best-if-Used-by	Recom-mended Home Storage Temperature

TABLE 4.1 (Continued)

Food Product	**Mode of Deterioration** ***(assuming an intact package)***	**Critical Environmental Factors**	**Shelf-life** ***(average)***	**Date Most Suitable for Product**	**Additional Information**
Long Shelf-Life Foods					
Dehydrated Foods	Browning, Rancidity, Loss of Pigment, Loss of Texture, Loss of Nutrients	Moisture, Temperature, Light, Oxygen	Dehydrated Vegetables 3-15 Months/Dehydrated Meat 1-6 Months/Dried Fruit 1-24 Months	Sell-by or Best-if-Used-by	Recom-mended Home Storage Temperature
Nonfat Dry Milk	***Flavor Deterioration***, Loss of Solubilization *(caking)*, Lysine Loss	Moisture, Temperature	12 Months	Best-if-Used-by	Estimate of Shelf-Life Beyond Sell-by Date; Store in Cool, Dry Place
Breakfast Cereals	Rancidity, Loss of Crispness, Vitamin Loss, Particle Breakage	Moisture, Temperature, Rough Handling	6-18 Months	Best-if-Used-by or Sell-by	Recom-mended Storage Conditions
Pasta	Texture Changes, ***Staling***, Vitamin and Protein Loss	Too High or Low Moisture, Temperature	Pasta with Egg Solids 9-36 Months/ Macaroni and Spaghetti 24-48 Months	Best-if-Used-by	---
Frozen Concentrated Juices	Loss of Turbidity or Cloudiness, Yeast Growth, Loss of Vitamins, Loss of Color or Flavor	Temperature	18-30 Months	Sell-by or Best-if-Used-by	Month of High Quality Left in Home Storage
Frozen Fruits and Vegetables	Loss of Nutrients; Loss of Texture, ***Flavor***, Color; and Formation of Package Ice	Temperature	6-24 Months	Best-if-Used-by	Recom-mended Storage Conditions
Frozen Meats, Poultry, and Fish	***Rancidity***, Protein Denaturation, Color Change, Desiccation	Temperature	Beef 6-12 Months/ Veal 1-14 Months/ Pork 4-12 Months/ Fish 2-8 Months/ Lamb 6-16 Months	Best-if-Used-by	Recom-mended Storage Conditions
Frozen Convenience Foods	Rancidity in Meat Portions, Weeping and Curdling of Sauces, Loss of Flavor, Loss of Color	Oxygen, Temperature	6-12 Months	Best-if-Used-by	Recom-mended Storage Conditions

TABLE 4.1 (Continued)

Food Product	Mode of Deterioration *(assuming an intact package)*	Critical Environmental Factors	Shelf-life *(average)*	Date Most Suitable for Product	Additional Information
Canned Fruits and Vegetables	***Loss of Flavor***, Texture, Color, Nutrients	Temperature	12-36 Months	Best-if-Used-by	---
Coffee	***Rancidity***, Loss of Flavor and Odor	Oxygen	Ground, Roasted, Vacuum-Packed, 9 Months/Instant Coffee 18-36 Months	Best-if-Used-by	---
Tea	***Loss of Flavor***, Absorption of Foreign Odors	Moisture	18 Months	Best-if-Used-by	---

[1] This date applies only if the product is packaged prior to sale. If unpacked or sold in bulk prior to sale, this product is exempt from an open date.

[2] Sweet corn has a shelf-life of 4 to 8 days, and apples range from 3 to 8 months at proper temperature. For this specific information, see Theodore Labuza *et al.* "Open Shelf Dating of Foods," Dept. of Food Science and Nutrition, University of Minnesota, report prepared for the Office of Technology Assessment, 1978.

NOTE: When known, the primary mode of deterioration is in bold italic type.

(OTA 1979).

Because of the variable nature of fresh produce and how it is handled in distribution, it is generally not dated. However, the recent growth of the fresh pre-cut salad business (estimated to be greater than one billion dollars) makes some type of dating necessary. Pre-cut mixed products have a shorter shelf-life due to senescence and microbial decay, unless some of the above technologies are employed with controlled temperatures in distribution. Because of the short experience of grocers and consumers with these products, some type of open date would help to move product from production to table more rapidly.

Microbiological Decay. Microbiological decay is a major consideration when establishing an open date, especially for perishable foods, such as fresh bakery goods, fresh and ground meats, fresh poultry, fresh fish, dairy products, cured meats (e.g., hot dogs and bacon), pasteurized fruit juice drinks, fresh and pre-cut fruits and vegetables and salads, and extended shelf-life pasta products. Most deli items including salads are included in this category of spoilage. Preservation methods used to control or destroy spoilage microbes include the following:

- Decrease temperature to slow down growth
- Pasteurize followed by refrigeration
- Reduce moisture to slow down growth

- Add acid to lower pH and slow down or prevent growth
- Use CAP/MAP technology (including addition of carbon dioxide) which slows growth
- Incorporate active packaging technologies, such as oxygen scavengers that reduce microbial growth (e.g., in refrigerated pasta dishes)
- Use edible barriers or packaging that releases an antimicrobial agent

Temperatures below 45°F, a pH less than 4.5, and a reduction of moisture to a water activity less than or equal to 0.8 are usually sufficient conditions to prevent microbial growth. It is the responsibility of manufacturers to ensure absolute control over the presence of microbes, because some microorganisms are pathogenic to humans. Microbes such as *E. coli* in ground beef and *Salmonella* and *Campylobacter* in chicken can result in illness or death if consumed.

With the threat of microbes in food, consumers must be educated to use open dating as a guide for quality and *not* as a guarantee of safety (Labuza 1982). Proper use of open dating, control of temperature in distribution, and efficient stock rotation should ensure that consumers are purchasing goods of high quality that give them an adequate time of use in the home.

Chemical Deterioration. Tissue damage that occurs during food processing releases food chemical constituents. These constituents enter an atmosphere of cellular fluids and react with each other or with other external factors such as oxygen leading to food deterioration and decreased shelf-life. The following are the major reactions which can occur and cause a decrease in quality and nutrient levels (Labuza 1982).

Enzymatic. Damaged cells release enzymes which enhance deterioration. The browning of damaged produce, such as bananas and peaches, demonstrates enzymatic deterioration, which is enhanced at room temperature. The methods for retarding degradation are similar to senescence and microbial decay (Labuza 1982).

Lipid Oxidation. Lipid oxidation is a mode of deterioration for fried snacks, nuts, dried meats, dried vegetables, dried fish, dried poultry, some dairy products, semi-moist meat products, pre-cooked refrigerated meats and fish, cured meat and fish, coffee, cooking and salad oils, margarine, most cereals, and spices. The process results from oxygen attacking unsaturated fats and can cause rancid off-flavors (such as occurs with potato chips), color changes (such as bleached dry vegetables), decreased quality, and the production of toxic substances. Lipid oxidation can be controlled by manipulating oxygen levels or by adding antioxidants (BHA, BHT, and EDTA). The oxygen permeability of the packaging material must be considered in determining an open date. Handling instructions for the consumer should be provided on the extent and control of such reactions to better maintain the quality of the food (Labuza 1982).

Non-Enzymatic Browning (NEB). NEB is the result of certain sugars reacting with proteins during processing and storage. It can cause darkening, off-flavors, and a decrease in protein nutritional quality in products such as powdered dairy products, dry eggs, dry drink juice mixes, semi-dry meats and fish, non-acid canned goods, breakfast cereals, cake mixes, fortified pastas, semi-moist breakfast bars, and frozen concentrated juices. In the latter, vitamin C degradation also leads to unacceptable darkening. Both storage temperature and humidity are critical with respect to external relative humidity. This affects the rate of moisture permeating through packaging for dry goods and the rate of browning, therefore, affecting shelf-life. For the most part, manufacturers of these longer shelf-life products either put no date on their products, or if they do, they use a "best if used by" date and assume some average distribution condition (and hope that most products are consumed before that date rather than sitting on the shelf). With canned goods, because the product is in a hermetically sealed container, which is impermeable to oxygen, moisture, and light, the major mode of deterioration is browning which is very slow at typical room storage temperature. Thus, some manufacturers date such products three years beyond the production date (Labuza 1982). Most likely, this date has never been verified in testing but is likely to be the limit of consumer storage.

Other Chemical Reactions. Other chemical reactions affecting shelf-life of food products include loss of vitamins and light oxidation of pigments. As with the other chemical reactions, temperature, oxygen level, moisture content and light are all important factors to consider in predicting shelf-life (Labuza 1982).

Physical Degradation.

Physical Bruising/Crushing. Although not equated into shelf-life determinations, physical abuse of a food (especially of fruits and vegetables) leads to microbial growth and product decay, because of the damage to cells allowing for invasion by spoilage bacteria.

Wilting. The problem of wilting is most prevalent in fresh leafy vegetables (whole or pre-cut) and tuber vegetables. It results in decreased crispness or an increase in the rate of senescence reactions. Good moisture barrier packaging can prevent wilting in most refrigerators, which have a relative humidity lower than 50% and cause moist products to dry-out. Too high a barrier to moisture loss can lead to package fogging and subsequent microbial growth. If the fresh produce is unpackaged, the use of sprays (such as used during grocery display) can reduce wilting. Overall, though, no data are available to tie shelf-life to moisture loss because of all the variables.

Texture/Stickiness. Moisture levels have a major impact on products such as candy, semi-moist foods, cake and bread products. A decrease in moisture levels increases the hardening of semi-moist products, so high moisture barrier packaging is required. With high levels of external humidity, potato chips, dried or fried snacks and crackers become soft due to moisture gain (Labuza 1982). High humidity and temperature also cause foods and convenience dry meal mixes, instant coffee, instant tea, and dried drink mixes to become sticky and caked. In baked, high moisture wheat flour products, such as bread and cake, staling is an important mode of deterioration to consider. This is the result of starch/protein interaction. Unlike many other reactions, the rate of this reaction is actually increased with lower temperatures, thus bread stales faster when refrigerated than when at room temperature. Freezing bread, however, ceases the physical change. In order to predict the end of shelf-life, the moisture permeability of the package and the temperature/relative humidity conditions as a function of time during distribution must be considered (Labuza and Schmidl 1988). In general, most food processors do not have enough information to set a reliable date for such products, except perhaps for fried, baked or extruded snacks.

Temperature Induced Texture Changes. Many texture changes in food are caused by continuous temperature fluctuation. This can lead to a loss of quality in many foods. In frozen foods, for example, products lose tissue moisture and the rate of chemical reactions increases under continuous thawing and freezing conditions. Temperature fluctuation also leads to the undesirable formation of ice crystals in frozen dairy products, e.g., ice cream, and freezer burn or discoloration on the surface of some food, e.g., meats. Another effect of temperature fluctuations is evident in emulsified products, for example, salad dressing and mayonnaise (Labuza 1982). An emulsion is a homogenous mixture consisting of liquid droplets dispersed into another liquid in which it is immiscible (McWilliams 1993). The stability of such products is jeopardized if there is a lack of temperature control, resulting in the undesirable physical separation of the product. Obviously, without good data of the distribution conditions, dating of such products is a guess.

Accelerated Shelf-Life Testing (ASLT)

For perishable and some semi-perishable foods, manufacturers may be able to determine the products' shelf-life by testing samples for quality at regular time intervals. Such testing may not be feasible, though, for nonperishable food products that can have shelf-lives of over three years long. In order to test such products, accelerated shelf-life tests (ASLT) are necessary (OTA 1979).

ASLT involves measuring the rate of quality loss experienced by a food product at two or three constant environmental conditions. With such information, the product's rate of deterioration can be determined as a function of time and/or humidity (See OTA 1979 for details and examples of such models and mathematical equations). Note that constant temperature and humidity are pertinent, especially in foods where water activity (a_w) and moisture content play important roles in the food's degradative process (Taoukis and Labuza 1996).

Due to lack of time, money, and qualified scientific personnel in many food manufacturing companies, however, ASLTs are not frequently performed. Therefore, many open dates are no more than very good guesses or industry practice (OTA 1979; Labuza and Szybist 1999a). A good review of the mathematics of shelf-life testing is by Taoukis *et al.* (1997) while Labuza and Schmidl (1985) give some good examples.

CHAPTER 5

CURRENT PRACTICES

Open-Dating Terminology

While the wording of open dates is currently established by manufacturers, Table 5.1 lists commonly used open-dating terminology and their definitions.

TABLE 5.1
OPEN-DATING TERMINOLOGY

Open-Dating Terminology	*Definition*
Production Date or Pack Date	Historical meaning; gives the date on which the product was manufactured or put into the final package. Used on prepackaged fresh fruits and vegetables, where shelf-life depends on the freshness of the product when harvested.
Sell-by Date	Helps in stock rotation to get the products out so the consumer can purchase the product at a point which will still give them adequate time for home storage before the end-of-shelf-life. Printed dates are usually very good guesses or industry practice based on assumed conditions.
Best-if-Used-by Date (Sometimes called Better if Used By)	The estimated point where the product quality loss reaches a level still generally acceptable but after which it fails to meet the high quality standard. Ambiguous date as to when the product should be taken off the supermarket shelf and confusing for the stock rotators.
Combination Date	"Best if used within __ days of (date)." The "__ days of" part makes this phrase a "best-if-used-by" date, while the "date" given represents a "sell-by" date.
Use-by Date	Commonly interpreted as "it dies or you die if you eat it" (Labuza 1982). The date determined by manufacturers as the end of useful quality life of the product.
Freeze-by Date	Often on meat or poultry in conjunction with another date, such as a use-by date. Helpful to the consumer and helps the store in terms of product movement.
Closed or Coded Date	Numbers used by the industry that indicate production lots. May represent a packing date, but not written for the consumer to understand. Important number for product identity in case of recalls.

(OTA 1979; Labuza 1982; FSIS 1995a).

Supermarket Data

Current data on open-dating practices at the supermarket level were collected by the University of Minnesota Retail Food Industry Center from October 1997 to September 1998. The labeling information of all food products was recorded from a large supermarket on the outskirts of St. Paul, Minnesota. After receiving verbal permission from the manager on duty, the following data were collected from food labels: *Name of Product*, *Product*, *Company Name and Address*, *Dating System*, the actual *Date* written on the package, and any *Additional Comments* related to the open-date system.

It should be noted that in the State of Minnesota (as well as Massachusetts) open dates are only required on products with 90 days of shelf-life or less and that fresh fruits and vegetables, meat, and poultry are excluded (also see Chap. 6). Since these are the most stringent dating requirements in this country currently, any dating on products with greater than 90 days of shelf-life is being done voluntarily by the manufacturer.

The data from this study were entered into Microsoft Excel 4.0. The following part of this chapter discusses the data in some detail. The food in these sections was grouped together as they were found in the grocery store aisles. From this data, Table 5.2 was created to summarize common open-dating practices seen at the retail level. The final column in the table demonstrates the open-dating system recommended, which is detailed in the proposed regulation (Chap. 7).

Produce Section. While most fresh fruits and vegetables do not carry an open date, there were several products in the produce section that did. Most precut fruits and vegetables, for example, bore a "use-by" or "best-if-used-by" date. On a bag of CAP/MAP spinach salad, however, there was no date on the package. The New Star Fresh Foods Company was called to question their decision not to date the product, but the phone number written on the package did not reach the California-based company; it was actually the number for a steel company in Chicago. Also regarding the CAP/MAP salad products, on October 21, 1997, a Salad Time baby spinach product was found to have a "best-if-used-by" date of "10 20 97". The product was brought to the attention of a supermarket employee, who responded that some out-of-date products which still seem fresh in visual appearance will be kept out on the shelf until the visual quality decreases or a new batch is received. While this may not be a safe practice regarding CAP/MAP foods, the employee informed us that it was a common practice.

TABLE 5.2
CURRENT OPEN-DATING PRACTICES AND RECOMMENDATIONS

Name of Food	Food Category	Common Open-Date Practice	Recommended form of open date
CAP/MAP fruits & vegetables	Refrigerated perishable	Use-by/ Best-if-used-by	Best-if-used-by & TTI
Deli meats (sliced in-store)	Refrigerated perishable	Sell-by	Best-if-used-by
Deli meats (prepackaged)	Refrigerated perishable	Sell-by/ Best-if-used-by	Best-if-used-by & TTI
Deli cheese & salads (packed in-store)	Refrigerated perishable	Sell-by	Best-if-used-by
Fresh meat & poultry	Refrigerated perishable	Sell-by/Use or Freeze-by	Best-if-used-by/ Freeze-by & TTI
Milk	Refrigerated perishable	Sell-by	Best-if-used-by & TTI
Milk products	Refrigerated perishable	Sell-by/Best-if-used-by/Use-by	Best-if-used-by & TTI
Juice	Refrigerated perishable	Sell-by/ Best-if-used-by	Best-if-used-by & TTI
Eggs	Refrigerated perishable	Expiration date	Best-if-used-by & TTI
CAP/MAP pasta	Refrigerated perishable/ semiperishable	Best-if-used-by/Use or Freeze-by	Best-if-used-by/ Use-by/Freeze-by & TTI
Refrigerated dough	Refrigerated perishable/ semiperishable	Best-if-used-by	Best-if-used-by/ Use-by
Yogurt	Refrigerated semiperishable	Sell-by/ Best-if-used-by	Best-if-used-by/ Use-by
Processed cheese (prepackaged)	Refrigerated semiperishable/ nonperishable	Best-if-used-by/ Sell-by/Expiration date/Use-by	Best-if-used-by/ Use-by
Butter, margarine, vegetable spread	Refrigerated nonperishable	Best-if-used-by	Best-if-used-by/ Use-by
Frozen meat & poultry	Frozen perishable/ semiperishable	Code date	Best-if-used-by/ Use-by & TTI

TABLE 5.2 (Continued)

Name of Food	**Food Category**	**Common Open-Date Practice**	**Recommended form of open date**
Frozen foods	Frozen nonperishable	Best-if-used-by/ Sell-by	Best-if-used-by/ Use-by
Fresh fruits & vegetables	Perishable	Pack date	Pack date
Bakery goods (in-store)	Perishable	Sell-by	Best-if-used-by
Potato chips	Perishable/ Semiperishable	Best-if-used-by/ Sell-by/Use-by	Best-if-used-by/ Use-by
Diet soda	Semiperishable	Best-if-used-by	Best-if-used-by/ Use-by
Canned vegetables	Nonperishable	(None)	Best-if-used-by/ Use-by
Salad dressing	Nonperishable	Best-if-used-by	Best-if-used-by/ Use-by
Shelf-stable sports drinks & juices	Nonperishable	Best-if-used-by/ (None)	Best-if-used-by/ Use-by
Canned fruit	Nonperishable	Best-if-used-by/ (None)	Best-if-used-by/ Use-by
Instant stuffing/potato	Nonperishable	Best-if-used-by	Best-if-used-by/ Use-by
Instant pasta meals	Nonperishable	Best-if-used-by	Best-if-used-by/ Use-by
Nonfat dry milk	Nonperishable	Sell-by	Best-if-used-by/ Use-by
Dry bakery ingredients	Nonperishable	Best-if-used-by/ Expiration date/ (None)	Best-if-used-by/ Use-by
Spices	Nonperishable	Best-if-used-by/ (None)	Best-if-used-by/ Use-by
Oil/shortening	Nonperishable	Best-if-used-by/ (None)	Best-if-used-by/ Use-by
Candy	Nonperishable	(None)	Best-if-used-by/ Use-by
Cookies/crackers	Nonperishable	Sell-by/(None)/Best-if-used-by/Use-by	Best-if-used-by/ Use-by
Coffee	Nonperishable	Best-if-used-by	Best-if-used-by/ Use-by
Tea	Nonperishable	Best-if-used-by	Best-if-used-by/ Use-by
Water	Nonperishable	Expiration date	Best-if-used-by/ Use-by

TABLE 5.2 (Continued)

Name of Food	Food Category	Common Open-Date Practice	Recommended form of open date
Soda	Nonperishable	Best-if-used-by	Best-if-used-by/ Use-by
Canned soup	Nonperishable	Best-if-used-by	Best-if-used-by/ Use-by
Instant noodles	Nonperishable	Best-if-used-by	Best-if-used-by/ Use-by
Cereal	Nonperishable	Best-if-used-by	Best-if-used-by/ Use-by
Baby food	Nonperishable	Best-if-used-by	Best-if-used-by/ Use-by
Baby formula	Nonperishable	Use-by	Use-by

Fruit dips, salad dressings, tofu, and wonton wrappers found in the refrigerated food section were also dated. There was little consistency with the dating of these products. Some companies chose to use either "sell-by" dates, "use-by" dates, "best-if-used-by" dates, "snack-by" dates, or only the date with no explanation, such as with Litehouse Veggie Dip products.

Deli and Bulk Foods. While many foods in the deli were dated when the product was purchased, i.e., lunch meats and salads, there were also several breads, spreads, and sandwich toppings that were dated by the manufacturer. Many products in this area, if dated, contained "sell-by" dates, including many of the breads and fresh uncooked pizzas. As for the products that were dated at the time of purchase, the date was printed by computer along with the net weight and price of the product. Most foods prepared, e.g., deli trays and sandwiches, had one to three day shelf-lives. Many of the ready-to-eat hot foods, including hot fried chicken, hot rotisserie chicken, hot entrees, and fried appetizers, only had a 30 minute to three hour "sell-by" date. From visual observation, however, many of the products were being offered for sale long after their acceptable shelf-lives.

Once packages of deli meats were opened and shaved/sliced, the products had "sell-by" dates between one to five days. Many of the cheese slices had about one week shelf-life, and salads were dated for three to seven days of shelf-life. Temperature control is important for the open dates on these products to have any validity, and since these products are not dated until there is a consumer purchase, the rotation practices of the deli employees are crucial.

As for the bulk food products in this area, it would be difficult to determine the shelf-lives because the consumer packages the products him/herself, and the bins are refilled before all of the older product is emptied out.

Meat Section. Just as in the deli, the meat which was packaged in-store was given a date at the time of packaging. The computer was programmed to print a specific shelf-life for specific products. Ground beef, for example, was given a "sell-by" date of 24 hours. Other meats ranged between a three-to six-day shelf-life. The practice was similar for seafood. As for the prepackaged meats, many of the frozen meats did not contain a date on the package. Most prepackaged refrigerated meats had a "sell-by" date, and some also had a "use or freeze-by" date or "packed on" date. The Oscar Mayer lunch/snack products were dated with a "use-by" date.

Milk and Milk Products. The milk section was observed on November 25, 1997, and ranged in dates printed from November 28 to mid-December. Most milk cartons were printed with "sell-by" dates, and some of them also included the following statement: "For best quality, use within 7 days of opening". Some of the non-dairy creamers used "sell-by" or "use-by" dates. Most of these products still had one to four months of shelf-life. The Reddi Whip cans of non-dairy whipped topping were the only products in this section to use an expiration date.

Other products in this area included ricotta cheese, sour cream, and cottage cheese. The open dates on these products varied. Some used "sell-by" dates, "use-by" dates, or "best-if-used-by" dates. Others printed an explanation similar to that seen on some of the milk cartons, but with a recommended storage temperature: "Quality assured 7 days beyond date on bottom if properly refrigerated (33-40)".

Table 5.3 shows all of the products which stated some sort of temperature recommendation. There were six different temperature-related statements in this section. Although most of the statements were similar in wording, there was a difference in the proper temperature ranges recommended. Some of the printed temperature ranges spanned over 11 degrees and exceeded the safety level of 41°F. Although most of these foods would not be a safety issue, suggesting storage conditions out-of-compliance with the recommended refrigerator temperature sends an improper mixed message. Because the deterioration of a product's quality is a function of temperature, there would be a significant difference in the shelf-life of a perishable product held at 33°F compared to the same food held at 44°F.

TABLE 5.3
CURRENT OPEN-DATING PRACTICES ON MILK PRODUCTS

MILK PRODUCTS			
Product & Address	***Open Date***	***Printed Dates***	***Additional Comments***
*(12 oz) **Old Home Dry Curd/ Cottage Cheese** Old Home Foods St. Paul, MN 55103	Use by	11/25/97	Quality assured through date on bottom if properly refrigerated (40-44).
(8 & 12 oz) **Chive-Onion Creamy Potato Topping**, (8 & 16 oz) **Fat Free Sour Cream**		DEC. 18 & DEC 23 DEC 18 & DEC 11	Quality assured 7 days beyond date on bottom if properly refrigerated (33-40).
(8oz) **Sour Cream with Chives**, (12 oz) **Light Potato Topping & Dip (Chive & Onion)**, (8 & 16 oz) **Sour Lean Light Sour Cream**, (8 & 16 & 32 oz) **Sour Cream** Old Home Foods, Inc. St. Paul, MN 55103	---	DEC 23 DEC 23 DEC 18 & DEC 23 JAN 02 & DEC 23 & DEC 23	Quality assured 7 days beyond date on side if properly refrigerated (40-44).
(12 & 24 & 32 oz) **Small Curd/Cottage Cheese Old Home Foods St. Paul, MN 55103	Use by	DEC 10/97 & DEC 10/97 & DEC 03/97	Fresh through date on bottom if properly refrigerated (40-44)
(5.5 oz) **Cottage Cheese (Mandarin Orange) (Peach) (Plain) (Pineapple)** Old Home Foods St. Paul, MN 55103		DEC 8 DEC 8 DEC 8 DEC 8	Fresh through date on lid if properly refrigerated (40-44)
(12 oz) **Cheddar Chive-Onion Sour Cream**, (12 oz) **Chive-Onion Sour Cream** Mid-America Dairymen, Inc. Springfield, MO 65802		DEC 28 DEC 28	Fresh through date on carton if properly refrigerated (33-44F)

* One of the few outdated products found at this particular store.
** There were five other Old Home cottage cheese products with the same explanation.

Therefore, if two containers of Mid-America sour cream were dated December 28 and one container was held at 33°F while the other was held at 44°F, the former product would actually have a longer shelf-life than its identical product stored at a higher temperature. The difference could be as much as two fold (Labuza and Szybist 1999a).

Refrigerated Foods. As shown in Chap. 2, the dating practices on yogurt containers varied among brands and sizes. There were even some inconsistent practices within the same manufacturing company. Recommended storage temperature information was printed on some of the yogurt containers, and in the case of Dannon yogurts, the products displayed a date without an explanatory statement. Instead, there was a hotline number printed on the container, which provided an explanation for the date. The company also offered to send the caller an informative pamphlet, which stated the following:

> ***Although the yogurt will remain fresh for at least a week beyond this date when properly refrigerated, its flavor changes and the yogurt becomes more tart with prolonged refrigeration.***

Although proper refrigeration was mentioned, there was no suggested temperature provided. Regardless, though, this method of dating food products is confusing and inconvenient to the consumers and the employees at the retail level responsible for rotating such products.

Concerning refrigerated juice, most containers still had almost a month to two months left of shelf-life according to their printed dates. "Sell-by" and "best-if-used-by" type of dates were most common. Some of the Tropicana and Dole products not only included an open date, but also included the following statement on the back of their packaging: "Best if used within 7 to 10 days after opening". A few juice products printed only the date on their containers without an explanation.

Most of the eggs had open dates. Many of the cheese products used "best-if-used-by" dates, but there were some "sell-by", "use-by" and expiration dates used on the cheese packages. On packages of Sargento crumbled blue cheese, the company stated that if the product was not used in three to five days after opening that it could be frozen for up to two months.

Due to modified atmospheric packaging conditions, DiGiorno pasta products had a one to three month shelf-life before their "use or freeze by" dates. Many refrigerated dough products also had about one to three more months until their printed dates and were labeled with "best-if-used-by" dates. As for the butter, margarine, and vegetable spread products, if there was a date on the container, most packages were dated with "best-if-used-by" type of dates; the dates averaged about one to seven months until their recommended end-of-use. There was a range of other refrigerated products from peanut butter and pudding to raisins and shredded potatoes. Most of these products had one to six months before reaching the date on its package. The date explanations varied from none to "sell-by/use-by" and "best-if-used-by" to "fresher before", "freshness guaranteed", and "wholesomeness great taste before" type of dates.

Bakery Goods. Most products in the bakery had fairly short shelf-lives, especially those products baked in-store. Breads and muffins, for example, had to be sold within three days. The "sell-by" dates on cakes and cookies were eight and ten days.

Other products sold in the bakery section included manufactured snack cakes such as Little Debbie, Hostess, and Tastycake products. Such foods had an average shelf-life of 30 days and used dating explanations such as "guaranteed fresh".

Frozen Foods. While freezing food products may extend shelf-life, it also affects the quality of the food. Dating frozen foods is important because long periods of time may result in foods with unappealing color and unacceptable taste and texture. Still, many frozen foods at the retail level were not dated.

With a variety of frozen pizza products on the market, only the Cub Foods self-rising pizza contained a "best-if-used-by" date. A few dough products were also dated. Campbell Soup Company was consistent and used "Recommend use by:" dates on their Swanson & Hungry-Man meals, Swanson breakfast entrees, Prego frozen lasagne, and Campbell's Restaurant Soup products. Pepperidge Farm's garlic bread, cake, and pastry products were all dated with "Sell-by" or "Recommended use by:" dates. Minute Maid juices were dated with "Best-if-used-by" dates. One of the few outdated products found in this study included a frozen Minute Maid concentrated juice. The product was observed in July 1998 and was dated November 16, 1997.

Feminine Hygiene, Shampoo/Conditioner, Lotions, Bath Oils, Hair Lotions. While there were no food products to be observed in this next section, several of the non-food items had open dates on their packages. Besides the over-the-counter medications, such as aspirin, many of the dandruff shampoos had an expiration date. According to personnel at the *Head and Shoulders* consumer affairs department, the product's expiration date was determined through shelf-life testing. Although no details were available concerning specific testing procedures, the active ingredient in the product was guaranteed effective for up to six months past the expiration date.

There was also a date on the bottom of *Banana Boat Sun Protection* products (suntan lotions). After calling the company's toll-free number, it was learned that the date represented the year of manufacture, and the product was guaranteed effective for three years past this date. The company's personnel had no further comment as to why this code was not more easily readable to the consumer.

Deodorant, Cough-Cold, First Aid, Vitamins, Diet, and Eye Care. Dr. Scholl's callus remover medication and several deodorant products had expiration dates. Many diet aid products were also dated and contained expiration or "use-by" dates. The Slim-Fast Foods Co. was not as consistent, though. Some of their diet aid products contained "best before" dates, "best use by" dates, or no explanation of the dates at all.

Toothpaste, Floral, Nylons, Candles, and Greeting Cards. As would be expected, there were very few dated products in this section. Only a few toothpastes and one of the mouthwash products contained expiration dates.

Institutional, Peanut Butter, School Supplies, Batteries, Pickles, and Jams-Jellies. Many of the institutional products did not carry an open date, e.g., dry foods, canned vegetables and fruits, condiments, and spices. Some of the pickle products had "best-if-used-by" dates. The dated pickles had about one year until their printed dates. As for jams, preserves, fruit spreads and jellies, some of these products, e.g., Smuckers' products, were open dated. Kraft Foods dated all of their jelly-type products but only included explanations for their Sunberry Farms products. Skippy, by Best Foods Division (CPC International Inc.), was the only peanut butter that was dated. Both the open-dated peanut butter and jellies had "best-if-used-by" type of dates.

There were also health foods in this section. Most soy milk cartons had about nine months to a year until their "best-if-used-by" type dates. Expiration dates were printed on some of the dried rice, noodles, and soups. There were "best-if-used-by" dates and "use-by" dates on most of the health bars and other shelf-stable health snacks. Some rice cakes offered for sale were very close to their "sell-by" dates.

Also of interest, Rayovac, Duracell, and Energizer batteries had "best-if-used-by" dates, and some of the batteries even had a form of time-temperature integrator (TTI); the batteries contained a color strip that faded over time to measure the amount of shelf-life left. The Fujifilm disposable camera and film also had an open date.

Condiments, Salad Dressing, Canned Vegetables, Kool-Aid, and Juice. A few of the canned vegetables were dated, such as Festal All Green Cut Spears of Asparagus and Heinz Premium Vegetarian Beans in Rich Tomato Sauce. The largest canned vegetable manufacturer in this case was Green Giant from the Pillsbury Company, which did not date its products. Salad dressings, on the other hand, were mostly dated with a "best-if-used-by" date as were mayonnaise products manufactured by Kraft, Hellmann's, Rich'n Smooth and Weight Watchers. Hot sauces and mustards were generally not dated except for those products manufactured by Kraft or Hellmann's.

Kraft, however, dated neither its powdered Kool-Aid drinks in packets and/or canisters, nor its Kool-Aid Kool Pop Freezer Bars. Twenty-fluid-ounce to one-gallon containers of the sports drinks produced by The Gatorade Co. (Gatorade) and Coca-Cola Co. (Powerade) were not dated. The sports drink manufactured by Pepsi-Cola, All Sport, was dated with the following explanation, "Drink by date on bottle for best taste." As for juices, all individual-sized tetra brik containers came with a "best-if-used-by" type of date, including the drinks from Minute Maid, a division of the Coca-Cola Co. Many canned juices, however, regardless of size, were not dated. This included Hi-C (a Minute Maid product), Hawaiian Punch (from Procter & Gamble), and Libby's (of Nestlé USA, Inc.). Welch's, Dole, and Campbell's companies dated all of their canned juices.

Prepared Foods, Potatoes, Canned Fruits, Rice-Beans, Ethnic Foods, Pasta-Sauces. Dole Packaged Foods Corp. and Oregon Fruit Products Co. were the only two canned fruit manufacturers to consistently date their products with "best-if-used-by" dates. Ocean Spray and Cub Foods also dated their cranberry sauce products.

In this section, some of the stuffings and flour tortillas had the shortest printed shelf-lives, and most were labeled with "sell-by" dates. Stove Top (Kraft Foods, Inc.) used a "Best when used by" date on some of its stuffing packages, and its products also seemed to have longer shelf-life dates compared to other dated stuffings. Other products in this section were more shelf-stable and were often dated with "best-if-used-by" type of dates if they were even dated at all. Betty Crocker (General Mills) was the only company to use an open date on its instant potatoes; the company was consistent in its dating practices. The Betty Crocker Suddenly Salad products and Tuna/Hamburger/Chicken Helper products all had "Better if used by" dates. Kraft was also consistent by dating all of its macaroni and cheese products.

Some canned foods in this section, such as stews and pasta, were also dated; consistency of dating depended more on the manufacturer than the actual product. "Best-if-used-by" type of dates were used on the following: Pasta Roni, by Golden Grain Co.; Campbell's, by Campbell Soup Co; and Franco-American, by Campbell Soup Co. Chef Boyardee and most of the Dinty Moore/Hormel canned stews and pastas did not have open dates.

Concerning the ethnic foods in this section, most of the spaghetti sauces were open dated, while the pasta was not. Most Asian sauces and dried noodles were not dated, but many of the Mexican beans and tortilla products were.

Sugar-Flour, Jello, Cake Mixes, Spices, Baking Chips, Oil-Shortening. While most products in this section had fairly long shelf-lives, two brands of nonfat dry milk used "sell-by" dates, which was commonly seen on perishable products, such as refrigerated milk. Both dry milk brands had almost a year until their printed dates. Many of the other dry ingredients contained only some sort of date without an explanation. Arm & Hammer Natural Baking Soda, on the other hand, had both a date and an explanation on its package. The only ingredient making up the product, however, was sodium bicarbonate; the functional properties of this substance are not affected over time under normal conditions unless it picks up moisture from the environment.

Kraft, which dated its dried pasta products and dried seasoning packets in the previous supermarket section, was not so consistent in regard to dating powdered gelatin and pudding mixes. In fact, out of all the Kraft products in this section, only the Jell-O No Bake Dessert Cherry Cheesecake and Instant Pistachio Pudding & Pie Filling were open dated.

Campbell's continued to be a consistent manufacturer in terms of using open dates on its products. Both Campbell's and Heinz dated their jars of gravy with "use-by" dates. The Quaker Oats Company used a "best before" date on its corn meal, and Nabisco's Planters peanut products used "best by" dates. "Better if used by" types of dates were also used on several of the prune/raisin packages.

The dry yeast products were all dated, but the open dates were not consistent. Some had expiration dates, "use-by" dates, or "guaranteed freshness" dates. Many bread machine mixes also had expiration dates. The McCormick Seasoning Blends (dried mixes) had open dates on all of their marinade and spice packets.

Pancake-Syrup, Popcorn, Fruit Snacks, Candy, Toast, Cookies-Crackers. Many of the dated products in this section contained "best-if-used-by" dates, for example, most pancake and waffle mixes, Quaker Oats Company (Aunt Jemima) syrup, and Knott's Berry Farms and Smuckers Fruit Syrups. Cereal bars by Hostess and Sunbelt were dated with "Last sale date" and "Guaranteed fresh" dates. Out of the popcorn products, only Jiffy Pop and Newman's Own Bags were dated. Very few of the candies had open dates, but European chocolates, RitterSport and Droste, were dated.

Most of the Jack Link Beef Jerky's had "sell-by" dates, but a few of their products contained only a date. The Slim Jim Brand of jerky also had "sell-by" dates printed on their packages. As for crackers and cookies, while Nabisco and Keebler products dominated this area, these products were dated with closed dates. Cub Foods cookies and crackers, Pepperidge Farms cookies and crackers, and Archway cookies were all open dated.

Coffee, Water, Tea, Cocoa. Some manufacturers chose to date coffee, tea, cocoa, and water. Most of the dates were "best before" dates. Nestlé, for example, used "best before" dates on many of their hot cocoa powders, but some of their cocoa products had only code dates. Cub Foods printed "use-by" dates on their cocoa powder. Nestlé Carnation non-dairy creamers had "best before" dates.

Celestial Seasonings, Inc. was the only company to date their boxes of tea bags. Most of the tea still had almost two years until the "best before end" dates. Lipton's instant ice tea powder was also dated. Just like many of the other Pepsi drinks, the products were printed with "Drink by date on bottle for best taste".

Several coffee products had "best before" dates printed on their containers. Dated instant coffees included Nestlé Hills Bros., Nestlé MJB, and Nicholas A. Papanicholas & Sons. Dannon, the same company that did not include an explanation for dates on their yogurt products, printed "EXP" with the date on their bottled water. Several other water bottling manufacturers used an expiration date, e.g., Ice Mountain, Crystal Geyser, Blue Mountain, and Evian.

Soup, Ramen Noodles, Paper Plates-Cups, Towels-Tablecloths, Canning Supplies, Bags-Wraps-Foils. Regarding canned soups, Campbell's was the most prominent manufacturer in this section. Most of the Campbell soup products still had about one to two years until reaching their "recommend use by date on lid" date. Cub Foods canned soups contained "best-if-used-by" dates.

Several of the dry soup mixes, such as Nile Spice, Near East, and Gourmet Award products, had "best-if-used-by" type of dates. Wyler's bouillon products by Borden Inc. used "best-if-used-by" or "use by" dates. Other dated products in this section included fruit pectin products by Kraft Foods, Inc. and Alltrista Co.

Cereal, Toys, Pop-tarts, Hot Cereals. Most cereal products had "best-if-used-by" dates, including cereals by Kraft, Preferred Products, Kellogg's, Quaker Oats, Malt-O-Meal Co. and General Mills. Kellogg's dated all of their cereals except variety paks and fun paks. Cereals from McKee Foods used a "guaranteed fresh" date.

Several of the other breakfast products also contained "best-if-used-by" dates on their hot cereals and toaster pastries. Even Nabisco chose to date their Cream of Wheat products. Most foods in this section still had six months to one year until their printed dates.

Beer, Canned Pop, Pop, Potato Chips, Bag Snacks, Tonic Water. With high consumer interest in freshness, several food companies have taken advantage of open-dating systems through advertisements. Since 1996, the message by Anheuser-Busch that "Fresh beer tastes better" has been highly publicized. The company uses a "Born-on Date" to boast their product's freshness. While several other breweries also insist that old beer tastes stale, many small breweries see open dating on beer as just a campaign by larger breweries to capitalize on the consumers' perception of "quality" (Anon. 1997b). O'Douls, a non-alcoholic brew by Anheuser-Busch, was the only other product in the beer section to be dated.

Regular sodas are shelf-stable, but diet sodas have a limited shelf-life due to aspartame (the artificial sweetener) degradation. Most diet beverages have an acceptable shelf-life of 90 days after manufacture when stored at room temperature, around 73°F (Stamp 1990). In the battle of the colas, Pepsi-Cola displayed an open date while Coca-Cola did not. "For best taste drink by date on bottle" was printed on all of the Pepsi-Cola products. The Pepsi-Cola Hotline Representatives claimed that the regular products are dated with a 90 day shelf-life, while the diet products are given 30 days. According to Table 5.4, the Pepsi drinks were all dated similarly with one to two months until their end-of-shelf-life, whether the product was regular or diet.

TABLE 5.4
OPEN DATING ON PEPSI-COLA PRODUCTS
(Collected 7/8/98)

PEPSI-COLA PRODUCTS *Pepsi-Cola* *Somers, NY 10589*		
Product	***Open Date***	***Date***
Pepsi	For best taste drink by date on bottle	SEP 28 98
Caffeine Free Pepsi	"	SEPT 21 98
Wild Cherry Pepsi	"	AUG 17 98
Diet Pepsi	"	SEP 21 98
Caffeine Free Diet Pepsi	"	SEP 21 98

Many snack foods were also dated. Crunch'n Munch and Cracker Jack toffee popcorn products were dated with "best-if-used-by" type of dates. Pepperidge Farms dated most of their cracker products with a "sell-by" date, although their 8 oz. cartons of goldfish snacks only contained a date. Many of the pretzels and chips had "best-if-used-by" type of dates, including the Frito-Lay foods. A few other chip manufacturers used "sell-by" and "use-by" dates on their packages.

Cigarettes, Specialty Breads, Hot Dog Buns, Hamburger Buns, Wheat Bread, White Bread. All bread products in this section were dated. Some of the dates included the month and the day (numerically), but some manufacturers printed an actual day (e.g., FRI or SAT) to represent the "sell-by" or "guaranteed fresh" date. Several packages only contained the day or the date without an explanation.

Baby Foods. All jars of baby food were dated with a "best-if-used-by" type of date. All infant formula, however, was dated with "use-by" dates as mandated by 21 CFR 107.20 (c).

Discussion and Conclusion. The information gathered from these supermarket observations clearly indicates that open dates in this country are often confusing, incomprehensible, and lack consumer-friendliness. Although the study searched specifically for open dates on food products, some of the dates were also difficult to locate. For example, on some of the canned vegetable juices by Campbell's, the date was written clearly on the top of the can. To understand what the date meant, however, the consumer would have to find the "recommend use-by" statement located within other statements on the container. There were several other examples where the date was not printed near the explanation. Other dates were hard-to-read because the print was too small or the ink smeared or the date was imprinted onto the package and difficult to see.

Dole was one of the few major manufacturers to consistently date its products with a "best-if-used-by" date. Campbell's and Pepperidge Farms also were fairly consistent with "recommended use by" and "sell-by" dates. General Mills and Quaker Oats usually dated their foods with "best-if-used-by" type of dates. Kraft dated all of its refrigerated products but only some of its nonperishable items. Nestlé was found to be inconsistent in dating only some of its foods. As for Nabisco and Pillsbury, many of their products were printed with only a code date, unless they were refrigerated.

Concerning the type of dates being used, some products were fairly consistent. Milk, for instance, generally had a "sell-by" date; however, explanations on many of the other refrigerated perishable products varied. Some manufacturers also chose to add storage information to their products. When used along with a consistent dating system, such information could be useful to the consumer.

As noted previously, most dating on non-refrigerated products was done voluntarily by the manufacturer, i.e., there was no requirement. Presumably, this was done because the company had consumer survey data which stated that such a practice would be very beneficial to the consumer. However, the great variety in type-of-date and inconsistencies in the dating practices among manufacturers makes the system ineffective. The consumer must be able to understand the dating system on food packages in order to use the dates properly. As evident by this study, the current status of the open-dating system in America is extremely inept and unable to efficiently serve its purposes.

CHAPTER 6

CURRENT REGULATIONS

Definition of "Food"

As defined in the Food, Drug and Cosmetic Act (as amended), the term "food" refers to the following: "(1) articles used for food or drink for man or other animals, (2) chewing gum, and (3) articles used for components of any such article" (21 USC 321f). Regulations pertaining to food are under the authority of the Food and Drug Administration (FDA), which is within the U.S. Department of Health and Human Services.

Fish is considered a food, but meat, poultry, whole eggs and alcohol are not included in this definition. Meat is covered under the Meat Inspection Act (21 USC 601 *et seq.*), poultry is under the Poultry Products Inspection Act (21 USC 451 *et seq.*), and egg inspection of whole eggs and egg breaking operations are under the Egg Products Inspection Act (21 USC 1031 *et seq.*); all of which are under the control of the U.S. Department of Agriculture (USDA). Alcohol is regulated by the Bureau of Alcohol, Tobacco and Firearms (BATF), which is a branch of the U.S. Department of Commerce.

Current Policies

Federal Regulations.

FDA Regulations. Open dating on food products is mandated in many countries, including the countries of the European Union, many South American countries, many of the Arabic States, the Scandinavian countries, Israel, Taiwan and Canada (Appendix B). As of 2001, the only federal regulations for open dating in the United States pertained to prescription and over-the-counter drugs and infant formulas. Under 21 CFR 107.20 (c), infant formulas must adhere to the following:

> ***A "Use by ____" date, the blank to be filled in with the month and year selected by the manufacturer, packer, or distributor of the infant formula on the basis of tests or other information showing that the infant formula, until that date, under the conditions of handling, storage, preparation, and use prescribed by label directions, will: (1) when consumed, contain not less than the quantity of each nutrient, as set forth on its label; and (2) otherwise be of an acceptable quality (e.g., pass through an ordinary bottle nipple).***

An article in a 1997 edition of *Supermarket News* stated that an increase in consumer pressure and media attention have strengthened the overall awareness of infant formula dating (Moore 1997). Depending on the volume of sales in a store, employees should be restocking these products on a daily to weekly basis and supposedly checking the expiration dates each time. According to the article, however, infant formulas within days before their expiration dates were discovered on grocery store shelves. There is no law regulating how far in advance of their expiration dates products must be removed from the shelves, but most grocery chains claimed to remove the products within a month beforehand. President and Executive Director of The Infant Formula Council in Atlanta, Georgia, Robert Gelardi, recommended checking the formulas on a monthly basis to ensure only high quality products. He also commented, "if a retailer inadvertently sells outdated infant formula to a customer, there are no serious health risks for the child. The effectiveness of the product is minimized, because vitamins begin to deteriorate after a certain period of time" (Moore 1997). In actuality, this is an inaccurate statement because there can be serious health risks if a young child is deprived of essential nutrients, which deteriorate over the whole storage time. Perhaps this practice will change, though, because in many stores infant formula is being kept under lock and key as this category has been the target of pilferage; the product is subsequently being sold off the street.

In late 1998, a bill was introduced into both chambers of U.S. Congress which would prevent state enforcement "relating to food safety warnings that are not identical to requirements in the Federal Food, Drug, and Cosmetic Act". In this bill (HR 4383 and S 2356), food safety issues that have been under the control of state regulations would continue to be exempt from any federal regulation. The exemption was specifically stated to include current open-dating policies. Food industry representatives believed that the bill would benefit the industry, as well as consumers, because warning labels determined only at the federal level would become more meaningful and consistent across the states (Dern 1998). The objective of this proposed law was to act as a means to make California Proposition 65 legislation illegal, i.e., an attempt to preempt state regulation of public health under the guise of inhibition of interstate commerce. Such uniformity across the state borders pertaining to open dating would benefit the industry and consumers in much the same way, but this bill goes against that principle of allowing states to regulate public health. As for the status of this bill, no action was taken by the end of the 105th Congress, and it was not re-introduced into the 106th Congress in 1999.

USDA Regulations. Meat, poultry and whole, uncracked eggs fall under the jurisdiction of the USDA. The Food Safety and Inspection Service (FSIS), one of USDAs public health agencies, has the responsibility of ensuring the safety, wholesomeness, and proper labeling and packaging of these products. Under the authority of this agency, an attempt to implement consumer involvement in food handling has begun. The following example is an illustration of a Safe Handling Label, which was started soon after a major outbreak of food poisoning in hamburgers resulted in the deaths of several children (FSIS 1993).

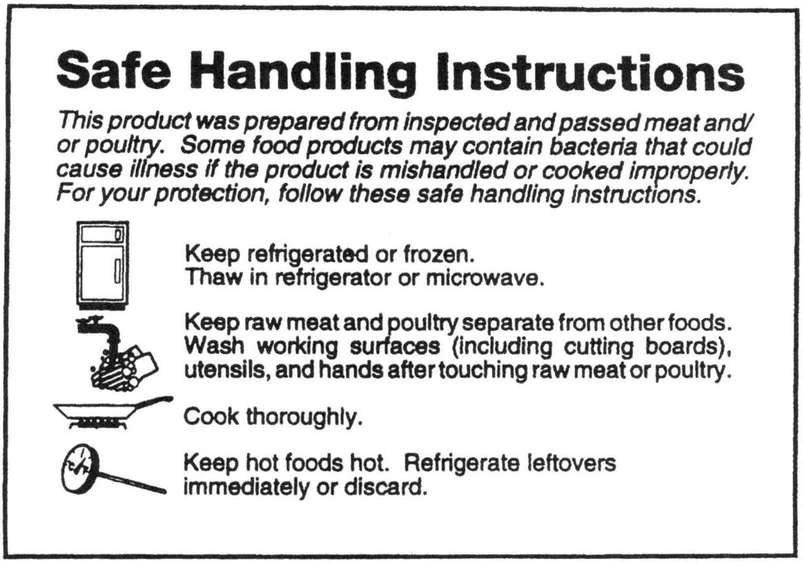

FIG. 6.1. USDA SAFE HANDLING LABEL

Although the federal government does not mandate an open-dating system on meat, if a date is printed on the package, it must be accompanied with an explanatory phrase, for example, "Full Freshness 10 Days Beyond the Date Displayed, When Stored at 40°F" (IFT 1981). The county of Los Angeles, California, is the only exception to this regulation. The county's local authorities have assured that their meat is under "a rigid local inspection program", therefore, an explanatory statement in that county is not required (USDA 1996).

According to FSIS, a retailer may sell meat and poultry products that are still wholesome even if they have gone beyond the expiration date on the label. It is illegal, however, for retailers to alter, change, or cover up the expired date with a new date.

National Institute of Standards and Technology (NIST). Established in the early 1900s, the National Institute of Standards and Technology (formerly known as the National Bureau of Standards (NBS)) is an agency of the U.S. Department of Commerce's Technology Administration (Anon. 1997c). According to Joan Koenig, Weights and Measures Coordinator at NIST, it is the responsibility of this agency to create uniformity among the states concerning local weights and measures laws, standards and practices.

In 1973, NBS devised the Model State Open Dating Regulations, which were later adopted by the National Conference on Weights and Measures (NCWM). NCWM is a "standards" writing organization consisting of federal government representatives, U.S. state and local weights and measures officials, industry representatives, and consumers. In these voluntary regulations, recommendations were made to use pull/sell-by dates for perishable/semiperishable foods with a spoilage risk within 60 days of packaging (IFT 1981).

NCWM continues to update its model systems regulations annually by contacting each state's weights and measures office, and NIST assists them by publishing this information. Open-dating regulations are found in NIST Handbook 130 (Appendix C). The handbook also contains a Uniform Regulation pertaining to open dating (representing an updated version of the 1973 Model State Open Dating Regulations). Here, NCWM states its concern about the lack of uniformity between state jurisdictions and the potential impediment on the orderly flow of commerce. The regulation suggests two options concerning open dating: (1) requiring open dating on all perishable foods, or (2) allowing for the voluntary usage of open dating of perishable foods as regulated by the Uniform Regulation (NIST 1997). It is interesting to note that this is the only regulation where a time period consisting of "at least one third of the approximate total shelf life of the perishable food" has been prescribed for determining the "sell-by" date.

In 1973, none of the 50 states had yet to adopt the NCWM regulations (IFT 1981). By 1997, NIST reported that Handbook 130 had been adopted in some form by 16 out of 53 U.S. states and territories. Table 6.1 summarizes the status of the NCWM Uniform Regulation in the each state/territory with a "YES/NO" type of answer in the columns labeled "LAW." Note that there are three styles used for the word "YES" and two styles for the word "NO" to indicate different levels of state participation.

The table shows that just 16 states mandate some form of open dating under the NCWM code, and only New Hampshire, Oklahoma, and West Virginia fully comply with NCWM open-dating regulations and its annual updates. However, when Joan Koenig was questioned about the status of Oklahoma, she verified that the state does follow the model regulations, but it has selected the second option of voluntary usage of standard regulations. She also noted that the state does *not* update its open-dating regulations annually; therefore, Oklahoma should have indicated a lower case "Yes" in the chart.

TABLE 6.1
STATUS OF ADOPTION OF NCWM OPEN-DATING STANDARDS BY STATE

STATE	LAW	STATE	LAW
Alabama	NO	**Nebraska**	NO
Alaska	NO	**Nevada**	NO
Arizona	NO	**New Hampshire**	**YES**
Arkansas	Yes	**New Jersey**	NO
California	NO	**New Mexico**	Yes*
Colorado	NO	**New York**	NO
Connecticut	Yes*	**North Carolina**	NO
Delaware	*No*	**North Dakota**	NO
District of Columbia	Yes*	**Ohio**	NO
Florida	Yes*	**Oklahoma**	**YES**
Georgia	Yes*	**Oregon**	Yes*
Hawaii	NO	**Pennsylvania**	NO
Idaho	NO	**Puerto Rico**	Yes*
Illinois	NO	**Rhode Island**	Yes*
Indiana	NO	**South Carolina**	NO
Iowa	NO	**South Dakota**	NO
Kansas	NO	**Tennessee**	NO
Kentucky	NO	**Texas**	NO
Louisiana	NO	**Utah**	NO
Maine	NO	**Vermont**	NO
Maryland	Yes*	**Virginia**	NO
Massachusetts	NO	**Virgin Islands**	Yes
Michigan	Yes	**Washington**	NO
Minnesota	Yes*	**West Virginia**	**YES**
Mississippi	NO	**Wisconsin**	NO
Missouri	NO	**Wyoming**	NO
Montana	NO		

http://ts.nist.gov/ts/htdocs/230/235/stlaw.htm

"YES": The capitalized letters represent states which have fully adopted the NCWM regulations. The state regulations keep current with each edition of Handbook 130.

"Yes": This style represents states that have adopted an NCWM recommendation in whole or in part from a particular year. Updates in Handbook 130, however, are not automatically incorporated.

"Yes*": The asterisk next to the "Yes" represents states which have adopted an open-dating regulation, but their respective regulation is NOT based on NCWM standards. This tends to be the case for most states that adopted state open-dating legislation before the development of the NCWM model.

"NO": Such states do not regulate any form of open-dating regulations on a state level.

"*No*": Delaware was listed with an italicized "No". This indicates that there is no required open-dating regulation in this state, but the uniform regulation is used as a guide. If a manufacturer were to ask the Delaware Weights and Measures Office for guidance about open dating, the office would refer them to Handbook 130 containing the Uniform Open Dating Regulation.

State Regulations. In 1979, OTA published a summary of state-mandated open-dating regulations in the United States. At that time, 22 states practiced some sort of open-dating policy. While the NCWM contacted States' Weights and Measures officials, OTA contacted the appropriate department in each state that would handle open dating. Open dating can be controlled in states by the State Department of Health, Department of Agriculture, Department of Weights and Measures, Department of Commerce, etc.

This chapter contains an updated state-by-state summary of open-dating regulations. Data were collected by phoning state regulators listed in the Interstate Milk Shippers (IMS) List. Since 20 out of 22 of the open-dating regulations from the OTA data pertained to milk or milk products, the IMS List was used as the first step to contact state officials. Questions regarding open-dating regulations were often directed to the state's respective departments, such as public health or standard weights and measures. In 1998, 29 states and the District of Columbia regulated an open-dating policy (Labuza and Szybist 1999a) (Table 6.2).

In some cases, an open date is applied by the manufacturers strictly on a voluntary basis irrespective of where the product is shipped. Comments from some of the state officials from places with voluntary open-dating policies tended to defend the fact that the government is more concerned with food safety issues; open-dating concerns food quality and should be left to the food processor.

The left half of Table 6.2 contains data that were collected and published by OTA in 1979. The differences among states regarding "primary products" and "form of open date" were significant, but even more interesting were the differences among the same products. For example, in 1979 the law in New Mexico stated that fluid milk and cream containers "were to be labeled with a legible "sell-by" date not to exceed 14 days including the date of packaging for pasteurized products and 5 days for raw products." Maryland's law pertaining to pasteurized milk products also required a "sell by" date, but this date was to be "7 days after the day of pasteurization" (OTA 1979).

The updated data on the right half of Table 6.2 represents the open-dating regulations as they stood in 1997-1998. Just as in 1979, there were many differences among the states. Table 6.3 illustrates some of the recent differences in "sell-by" dates for milk products as they vary across state borders. The data show that New Mexico and Maryland still differ in their legislation with respect to the maximum shelf-life of milk. However, while New Mexico eliminated the maximum number of days of shelf-life, Maryland doubled its set amount to 14 days. A major discrepancy is also found between the regulations of New York City and Virginia. While New York City has a maximum "sell-by" date of 96 hours for milk, this same amount of time is the minimum number of hours to be placed on a milk container in Virginia.

TABLE 6.2
PAST AND CURRENT OPEN-DATING REGULATIONS BY STATE

	1979			1998	
State	*Primary Products*	*Form of Open Date*	*Effective Since About*	*Primary Products*	*Source*
Alabama	Dairy	Sell-by	1975	Potentially Hazardous Foods	Department of Agriculture and Industries Agricultural Chemistry: ss. 80-1-22-.28
Alaska				Voluntary	
Arizona				Eggs	ALIS Online: ss. 3-719(E) (http://www.azleg.state.az.us/cgi-bin/wais)
Arkansas				Voluntary	
California	Dairy	Sell-by	1973	Milk and Milk Products	California Annotated Codes (1986): Division 15, ss.36004
Colorado				Voluntary	
Connecticut	Milk	Sell-by	1973	Milk and Milk Products	Connecticut General Statutes Annotated (1985): ss. 22-197b
Delaware				Voluntary	
District of Columbia	Perishable Products	Sell-by	1974	Perishable Products	District of Columbia Municipal Regulations (1997): Title 23, ss.2505
Florida	Dairy	Sell-by	1976	Milk and Milk Products	Florida Statutes Annotated: ss 502.042
Georgia	Milk and Eggs	Sell-by	1973	Food Products in Package Form	Rules of Georgia Department of Agriculture: ss. 40-7-1-.26
Hawaii				Milk and Milk Products	Hawaii Administration Rule: ss. 11-15-39 (c)
Idaho				Voluntary	
Illinois				Voluntary	
Indiana				Voluntary	
Iowa				Refrigerated Foods in Reduced Oxygen Packages	Iowa Code (1997): 481-31.11 (137A)

TABLE 6.2 (Continued)

	1979			1998	
State	*Primary Products*	*Form of Open Date*	*Effective Since About*	*Primary Products*	*Source*
Kansas				Voluntary	
Kentucky				Milk and Milk Products	Department for Health Services - Open Dating Requirements for Milk Products (902 KAR 50:080)
Louisiana				Voluntary	
Maine				Voluntary	
Maryland	Milk	Sell-by	1971	Milk Products	Annotated Code of MD (1996) (1997 Supplement): ss. 21-426
Massachusetts	Perishable & Long Shelf Life	Sell-by or Use-by	1979	Perishable and Semi-Perishable Foods Products	Code of Massachusetts Regulations 105 CMR: 520.119
Michigan	Perishable Products	Sell-by	1969	Milk and Milk Products	MI Dept of Agriculture, Dairy Division, Regulation No. 408. (http:/ / www.state.mi.us/ execoff/admincode)
Minnesota	Perishable Products w/Shelf Life < 90 Days	Sell-by or Use-by	1973	Perishable Products with Shelf Life < 90 Days	Minnesota Statutes Annotated (1996): ss. 31.781 - 31.784
Mississippi				Eggs	Mississippi Egg Marketing Law and Regulations-Regulation 1
Missouri				Voluntary	
Montana				Fluid Milk and Grade A Milk Products	Administrative Rules of Montana: ss. 32.8.201 - 32.8.205
Nebraska	Eggs	Pack	---	Eggs & Refrigerated Reduced Oxygen Packaged Food	Food Establishment: Reduced Oxygen Packaging Method: ss. 81-2,272.27. (http:// www.agr.state.ne.us) & Nebraska Graded Egg Act and Rules & Regulations

TABLE 6.2 (Continued)

	1979			1998	
State	***Primary Products***	***Form of Open Date***	***Effective Since About***	***Primary Products***	***Source***
Nevada	Dairy	Sell-by	1973	Voluntary	
New Hampshire	Cream	Use-by	1973	Refrigerated Prewrapped Sandwich & Prepackaged Perishable Foods (NIST)	NH Department of Agriculture, Markets & Food: Agriculture ss. 1413.04 and NIST Handbook
New Jersey	Dairy	Sell-by	---	Fluid Milk	New Jersey State Department of Health (1994): ss.8:21-10.1
New Mexico	Milk	Sell-by	1977	Milk and Milk Products	http://www.nmdaweb.nmsu.edu
New York				NYC - Milk and Milk Products	NYC Health Code: ss. 111.33
North Carolina				Smoked Fish	NC Administrative Code: ss. .0507e(5)
North Dakota				Voluntary	
Ohio	Perishable Products	Sell-by	1977	Perishable Products	OH Department of Agriculture (Regulation 901:3-57-04)
Oklahoma	Meat, Eggs	Sell-by	---	Voluntary	
Oregon	Perishable Products	Pack or Sell-by	1975	Packaged Perishable Products	Title 49: Chapter 616, ss. 616.815 - 616.830
Pennsylvania	Milk	Sell-by	1975	Milk	PA Code: Title 7, Part III: ss. 59.22
Rhode Island				Voluntary	
South Carolina				Voluntary	
South Dakota				Eggs	South Dakota Dept. of Agriculture: Law 39-11: ss.12:26:10
Tennessee				Voluntary	
Texas				Voluntary	
Utah				Voluntary	

TABLE 6.2 (Continued)

	1979			1998	
State	***Primary Products***	***Form of Open Date***	***Effec-tive Since About***	***Primary Products***	***Source***
Vermont				Voluntary	
Virginia	Dairy and Infant Formula	Sell-by	1974	Milk and Milk Products	ss. 2VAC5-490-40 (http://leg1. state.va.us/000/ reg.TOC02005. HTM#C0490)
Washington	Dairy and Others	Sell-by	1974	Perishable Packaged Foods	RCW 69.04.905
West Virginia				Perishable Products	West Virginia Code (1996):ss. 47-1-9
Wisconsin	Smoked Fish	Pack	1971	Smoked Fish	Wisconsin Administrative Code (1998): ATCP 70.22
Wyoming				Voluntary	

Sources for each state's data are stated in Appendix D

TABLE 6.3
OPEN DATING ON PASTEURIZED MILK PRODUCTS

State	Sell-by/Pull Date
California	Established by the processors/manufacturers
Connecticut	12 day maximum
Florida	Established by the processors/manufacturers
Hawaii	Established by the processors/manufacturers
Kentucky	Established by the processors/manufacturers
Maryland	14 day maximum
Michigan	Established by the processors/manufacturers
Montana	12 day maximum
New Jersey	Established by the processors/manufacturers
New Mexico	Established by the processors/manufacturers
New York City	96 hour maximum (pasteurized milk); 15 day maximum (ultra-pasteurized milk)
Pennsylvania	14 day maximum
Virginia	96 hour minimum; maximum number of days established by the processors/manufacturers

Sources for each state's data are stated in Appendix D

Appendix D contains extensive information on each state's current open-dating regulations. Massachusetts and Washington's statutes demonstrate two of the more stringent open-dating regulations in this country. The following excerpts describe *perishable food* and *semi-perishable food* as defined according to The Commonwealth of Massachusetts 105 CMR 520.11 (c):

> ***Perishable Food:*** ***A food product having an estimated shelf life of 60 days or less.***
>
> ***Semi-Perishable Food:*** ***A food product having an estimated shelf life greater than 60 days but less than 90 days.***

The following section of the regulation mandates that such food products contain an open date (105 CMR 520.119 (d)):

> ***Open Dating of Perishable and Semi-Perishable Food Products.*** ***No person shall sell, offer for sale, or have in his possession with intent to sell, prepackaged perishable or semi-perishable food products unless they are identified with a "sell-by date" or a "best if used by date" determined by manufacturer, processor, packer, repacker, retailer, or other person who had packaged such food products and displayed in the form specified in 105 CMR 520.119 (G).***

The regulations of Washington also require open dating on perishable-packaged food, but its definition applies to perishable-packaged food with a shelf-life of only thirty days or less:

> ***All perishable packaged food goods with a projected shelf life of thirty days or less, which are offered for sale to the public after January 1, 1974 shall state on the package the pull date (RCW 69.04.905 Perishable packaged food — Pull date labeling — Required).***

The definition of "perishable-packaged food goods" as defined in RCW 69.04.900(1) is as follows:

> ***...includes all foods and beverages, except alcoholic beverages, frozen foods, fresh meat, poultry and fish and a raw agricultural commodity as defined in this chapter, intended for human consumption which are canned, bottled, or packaged other than at the time and point of retail sale, which have a high risk of spoilage within a period of thirty days, and as determined by the director of the***

department of agriculture (sic) by rule and regulation to be perishable.

Irrespective of New Hampshire and West Virginia, which both mandate the open-dating regulations published by NIST, no other states have exactly the same regulations. Standards for the products to be dated, the form of the date, the maximum length of shelf-life, and definitions of key words (such as "perishable foods") do not hold the same over state borders when pertaining to an open date. The variation among states has a significant impact on both producers in interstate commerce and supermarkets with stores in different states, creating a significant problem in dealing with consumer satisfaction.

European Union (EU). In 1972, dedicated countries of the European Community committed themselves to achieve an 'ever closer union among the European peoples'. Their aim was to create the European Union (EU) to integrate and strengthen the political and economic communities of its members (Pryce 1987).

Within the EU legislation, open dating (or durability dating) was amended in Directive 97/4/EEC of the European Parliament and of the Council; it is found in Article 9 of 79/112/EEC. Article 9 mandates the use of "Best before" and "Use by" dates as shown in Appendix E. The following are excerpts of the EU open-dating legislation:

1. ***The date of minimum durability of a foodstuff shall be the date until which the foodstuff retains its specific properties when properly stored. It shall be indicated in accordance with the provisions of this article.***

2. ***The date shall be preceded by the words:***
   --"Best before..." when the date includes an indication of the day,
   --"Best before end..." in other cases.

The following excerpt was taken from Article 9a:

1. ***In the case of foodstuffs which, from the microbiological point of view, are highly perishable and are therefore likely after a short period to constitute an immediate danger to human health, the date of minimum durability shall be replaced by the "use by" date.***

Thus in the EU, the use-by date (at least for refrigerated perishable foods) has a specific food safety meaning, which is something not considered in the U.S.

Uniform legislation among the EU members simplifies food regulations across the continent's borders. It also allows the European people to understand the dates on their food products in any supermarket chain in any country of the Union. This increased confidence in an open-dating system is a luxury yet to be afforded by the American people across this country's state borders.

CHAPTER 7

PROPOSED REGULATIONS

Legislative Proposal for Minnesota

An initial attempt to improve open-dating legislation took place at the state level. The goal of passing such legislation would be important as a precedent for proposing a federal mandatory open-dating regulation.

Current Legislation. Minnesota currently has one of the most stringent open-dating laws in the country mandating that open dates appear on foods with a shelf-life of 90 days or less. Open dating is listed as "Quality Assurance Dating" in the Minnesota Statutes and states the following:

> ***Perishable foods which bear a quality assurance date of 90 days or less from the date of packaging shall be dated in accordance with the rules adopted pursuant to sections 31.781 to 31.789* (Minnesota Statutes, 31.7783, Subdivision 2).**

Perishable is defined as the following under 31.782 Definitions Subd. 3.:

> ***"Perishable food" means any food intended for human consumption (other than meat and poultry, frozen food, or fresh fruit or vegetables), which has a quality assurance date.***

Legislative Proposal. The legislative proposal prepared to update Minnesota's "Quality Assurance Dating" legislation was entitled "Redefining 'Perishable food' in Minnesota". The purpose was to redefine "perishable food" so that it included all reduced oxygen packaged food.

The current definition of "perishable food" is outdated in Minnesota due to advances in controlled atmospheric/modified atmospheric packaging (CAP/MAP) techniques in the food industry. Fresh fruits and vegetables, for example, are currently excluded from the "perishable food" category because of the variable nature of fresh products and how they are handled during distribution and storage. The recent growth of the fresh pre-cut salad business, however, forces the present logic regarding Minnesota's "perishable food" status, to be reevaluated. Pre-cut salad products require the CAP/MAP technology to extend shelf-life and prevent premature senescence and microbial decay. The same technology has also been applied to certain meat products.

Due to the inexperience of consumers and grocers in handling these products, a "quality assurance date" is necessary to help move the product from production to table more rapidly. Such dating is urgent to encourage proper rotation of these products and to help avoid accelerated deterioration due to retailers and consumers keeping their refrigerator temperatures higher than recommended. Rapid deterioration of the prepackaged foods at abused temperatures could lead to pathogenic growth, e.g., *Clostridium botulinum*, and could potentially result in human fatalities.

To help promote efficient rotation practices of CAP/MAP products, these foods must be considered "perishable food" in Minnesota legislation. The following modification in the definition of "perishable food" was strongly recommended:

Subd. 3. "Perishable food" means any food intended for human consumption which has a quality assurance date. This definition excludes meat and poultry or fresh fruit or vegetables unless packaged under controlled atmospheric/modified atmospheric conditions. Frozen foods are also excluded.

A copy of the actual legislative proposal is shown in Appendix F.

Legislative Visit. A meeting was set-up with Representative Mary Jo McGuire (D) at the State Office Building (St. Paul, MN) on February 12, 1999, to discuss the following key points:

- To familiarize Rep. McGuire on the federal open-dating system (or lack of)
- To describe present research (this current study in particular) concerning open dating
- To familiarize the representative about personal concerns regarding the present wording of the open-dating (or quality assurance dating) definition in the Minnesota Statutes
- To ask for her support in updating this section of the Minnesota Statutes

Representative McGuire's intern, Jessica, was also present during the meeting, which lasted approximately 15 minutes. After discussing the key points of the issue, Rep. McGuire stated that she was extremely busy at the present time, but that she was very interested in the issue at hand. She handed the proposal to her intern and said that they would have the House Researchers look at it. If the House Researchers felt that a change in the wording of the legislation was valid, then the committee would be in contact. Before the meeting concluded, a final comment was made in regard to the overall section on "Quality Assurance Dating", and how it should be written more clearly. The representative made note of the comment for the researchers to consider.

Follow-up to the Legislative Visit. Following the meeting with Rep. McGuire, a thank you letter was sent. After two weeks, there was still no response from the representative regarding the proposed legislation, so then an e-mail message was sent. Two days later, the intern, Jessica, responded to the e-mail and stated that a House Researcher drafted the legislation, and Rep. McGuire would probably be the chief author.

Soon afterwards, the bill was officially introduced into the House of Representatives as bill number 1894. Representative McGuire's office suggested Senator John Marty as the senate author. Jessica wrote that the Senator would probably take the bill, but first he would amend it to get rid of the double negative. After several e-mails and phone calls to his office, however, Senator Marty seemed to be skeptical about the importance of the matter.

The bill was also referred to the Committee on Agriculture Policy. All of the committee members (Table 7.1) were e-mailed and asked to support the bill in the legislature. Several committee members replied to the e-mail message with questions or with interest. As of June 2001, no further progress on the bill was made.

TABLE 7.1
MINNESOTA STATE COMMITTEE ON AGRICULTURE POLICY
(March 1999)

Chair: Rep. T. Finseth	
Vice Chair: Rep. T. Westrom	
DFL Lead: Rep. S. Wenzel	
Rep. G. Cassell	Rep. Rostberg
Rep. G. Davids	Rep. L. Schumacher
Rep. D. Dorman	Rep. R. Skoe
Rep. B. Gunther	Rep. Storm
Rep. E. Harder	Rep. H. Swenson
Rep. A. Juhnke	Rep. Tunheim
Rep. G. Kubly	Rep. Wejcman
Rep. R. Ness	Rep. Westfall
Rep. M. Otremba	Rep. T. Winter
Rep. D. Peterson	

http://www.house.leg.state.mn.us/hinfo/e-mail.htm

National Uniform Food Safety Labeling Act

Congresswoman Nita Lowey (R-NY) is a strong supporter of a mandatory open-dating legislation and in 1999 introduced open-dating legislation into U.S. Congress. The National Uniform Food Safety Labeling Act (HR 1346) was introduced into the House of Representatives on March 25, 1999, to "amend the Federal Food, Drug, and Cosmetic Act to safeguard public health and provide to consumers food that is safe, unadulterated, and honestly presented." Section 6 of the proposal is entitled "Freshness date" and states the following:

> ***(x)(1) Unless its label or labeling bears the date upon which the food should no longer be sold because of diminution of quality, nutrient availability, or safety. The freshness date shall be stated in terms of the day and month of the year if the food will not be fresh after 3 months on the shelf, or in terms of the month and year if the product will be fresh for more than 3 months on the shelf. The phrase 'use by' shall precede the date.***

Proposed Federal Regulation

In the U.S., 29 states and the District of Columbia regulate open dating, but without unifying principles it is confusing to use for storeowners (in terms of stock rotation) and for consumers. To eliminate the confusion surrounding open dates, a proposed federal regulation was created.

The proposed regulation was prepared from the information presented in this study, and the style was modified from several current open-dating regulations, e.g., Code of Massachusetts Regulations. The regulation would be added to 21 CFR 101 Food Labeling, and its purpose would be to mandate uniform open dates in the United States. The open-dating proposal contains background information, current policies, benefits, public health importance, economic impact, environmental impact, risk analysis, and the recommended use of TTIs. Stakeholders were identified as being everyone who buys market food in this country. However, the food industry and the government would be the biggest stakeholders because the passing of a mandatory open-dating system would possibly require the financial responsibility of conducting shelf-life tests for the food industry and the monitoring or setting of standards of open shelf-life determinations by the government. The actual proposal is found in Appendix G.

The proposal was sent to two consumer groups for support: Dr. Mike Jacobson at the Center for Science in the Public Interest (CSPI) (Washington, DC), and Ms. Nancy Donley of Safe Tables Our Priority (STOP). In addition, there was communication with Congresswoman Nita Lowey (R-NY), Senator Paul Wellstone (D-MN) and Congressman Bruce Vento (D-MN). As of July 2001, no action occurred in Congress indicating that this issue is not considered important.

CHAPTER 8

JUDICIAL ACTION

The Tenth Amendment

The Tenth Amendment of the U.S. Constitution gives the states power over areas not covered in the Constitution, i.e., "The powers not delegated to the United States by the Constitution nor prohibited by it to the States, are reserved to the States respectively, or to the people". This specifically allows states to mandate laws related to public health, e.g., open dating. Due to costs, as well as to enhance commerce between the states, however, most states have delegated such regulation to the federal government. This delegation does not imply that state action is not warranted. The State of California Proposition 65 is one example. This law requires a warning label on products that contain potential teratogens and carcinogens. The state has such authority under the Tenth Amendment.

Grocery Manufacturers of America v. Department of Public Health of Massachusetts

Since the federal government does not regulate any type of open-dating regulations on food besides infant formula, such policies have been created by some of the states by authority of The Constitution. In Grocery Manufacturers of America (GMA) Incorporation versus the Department of Public Health (393 NE 2d 881, 1979), an argument went to the Supreme Judicial Court of Massachusetts to challenge the validity of an open-dating policy mandated by the Massachusetts Department of Health. The Massachusetts policy required food vendors to disclose a "last date of use" or "pull date" on all food products. While GMA, which is an association of food manufacturers, insisted that the legislation imposed an improper burden on interstate commerce because the manufacturers in other states did not have such stringent regulations, the court ruled in favor of the state's authority. It held that "evidence failed to establish that operation of regulation would impose impermissible burden on interstate commerce" and that the open-dating regulation "bore reasonable relation to goal (sic) of consumer protection" (393 NE 2d 881, 1979). Since the Massachusetts Department of Health held that out-of-date products have an increased risk of non-safe agents, the stricter state law related to health, had power granted under the Tenth Amendment. While this is the only federal court case related to the open dating of food products, it substantiates the ability of states to mandate such a practice.

Domestic Regulatory Penalties

The State of Alabama has recently adopted the "Alabama Safe Foods Act of 2000" which describes penalties for out-of-date, misbranded and adulterated foods (Anon. 2000) (see Appendix H). The Act defines several "Classes" of violations for three primary areas of food (meat, dairy, and baby food/infant formula). The classification of violation is determined by the number of out-of-date products and the amount of time past the open date. The fines and actions for violating the act range from a written warning (for a Class I violation) to a maximum $10,000 penalty and the possible revocation of one's food safety permit (for a Class V violation). According to the Act, misbranding refers to the "Obscuring, removing, or extending existing open date statements" and shall be considered a Class IV violation. The Act is the most aggressive action taken by a state at this time to hold persons responsible for out-of-date foods. New Hampshire and Washington also have monetary penalties up to $1000 and $500 respectively for not complying with open-dating regulations.

Legal Action Abroad

Legal action most likely has been taken against open-date offenders at the state level. It is difficult to determine how often this occurs, though, since most cases do not go past the local court systems, and there is very little published documentation. There is limited documentation, however, which demonstrates that questionable open-dating practices are not just a problem in the United States.

In one example, a Sainsbury store (a large supermarket chain in the United Kingdom) was charged for unethical open-dating practices. The store was fined $14,000 for selling foods past their "use-by" dates and trying to conceal the original open date with a new sticker. The food items included crab, beef burgers, ostrich goujons and chocolate roulade. The issue was brought to the attention of the authorities by a consumer who noticed the October 29 "use-by" label on the crab meat that he purchased covering another label stating October 27 as the "use-by" date. The Sainsbury store admitted to the offense, but stressed that all of the products would have been safe to eat and would not have invoked any potential threats (Butler 1998). This may be true, but it is not the key point of open dating; this defense only confuses the matter.

In another example (Besfamille 1998), the "Direction Generale de la Concommation et de la Repression des Fraude", the French public agency in charge of the quality of goods sold to the consumer, found 115 cases of relabeling of out-of-date meat during 1,200 inspections at the grocery store level. In France, meat must have a "use-by" date, which is set discretionally by the seller. Although no cases have been reported in legal literature, it is commonly felt by the consumer that this practice also occurs in the U.S.

In June 2001, the Canadian Food Inspection Agency fined Santa Maria Foods Corp. of Vancouver $100,000 for removing batch numbers and altering the "best if used by" dates on imported cheese (Anon. 2001). This was the second largest fine ever imposed by the agency.

Misbranded Food

An open date on a food package implies something to a consumer. To many of them the date indicates something about the shelf-life or safety of the food. To the producers it represents the time at which the loss of desired quality occurs. To the food retailer it tells them something about how fast to move the product to get it into the consumer's home before it spoils. The date also implies that the product is going to be stored properly, e.g., that the temperature in the refrigerated cabinet is maintained at 41°F or lower. If not maintained properly, the food may spoil before the date, leading to a disgruntled consumer or to a food poisoning incident. If no date is present, consumers may sort for those that are dated or become confused as to how long to store the food at home.

The presence or absence of a date has legal implications, with respect to being either misleading or misbranded. Specifically, as stated in Section 201(n) of the Food, Drug and Cosmetic Act, the main act that controls all foods except meat, poultry, eggs and egg breaking operations, misbranding includes the following:

> ***If an article is alleged to be misbranded because the labeling or advertising is misleading, then in determining whether the labeling or advertising is misleading, there shall be taken into account (among other things) not only representations made or suggested by statement, word, design, device, or any combination thereof, but also the extent to which the labeling or advertising fails to reveal facts material in the light of such representations or material with respect to consequences which may result from the use of the article to which the labeling or advertising relates under the conditions of use prescribed in the labeling...***

Although no similar wording appears in the Meat Inspection Act, Poultry Products Inspection Act, and the Egg Products Inspection Act, it is assumed that their respective products would follow the same principles. Currently for these latter products at the federal level, all labeling must have prior approval; however, this requirement was dropped in 2001.

False Labeling. False labeling on a food package is legally misbranding according to Section 403 (a) of the Food, Drug and Cosmetic Act:

A food shall be deemed to be misbranded —
If (1) its labeling is false or misleading in any particular...

According to 9 CFR 381, raw chicken cannot be labeled as "fresh" if the internal temperature goes below 26°F (FSIS 1995b). Many poultry processors were pre-freezing poultry to extend shelf-life, as well as to ensure a safer product by distributing to grocers under frozen conditions rather than increasing the risk of potential temperature abuse in refrigerated transport which is closer to conditions where pathogens might grow. However, after complaints by some consumer groups, USDA was forced to consider this practice of frozen distribution to be misleading to the consumer, and the product would therefore be misbranded. For this reason, some poultry processors now date their poultry with much shorter shelf-life dates. In late 1998, USDA issued a warning to a major poultry producer for still shipping frozen poultry and labeling it as "fresh".

Pertaining to eggs, FSIS has revised its regulations under the Egg Products Inspection Act (EPIA). It is now required that shell eggs have refrigerated transportation and storage temperatures of no greater than 45°F, including holding and display temperatures in the supermarket. This is to ensure against growth of a microbial pathogen in the egg, *Salmonella* enteritidis, which causes serious illness and is the most reported cause of illness from the *Salmonella* species. Although the cost of guaranteeing proper refrigeration may be considered too expensive to be feasibly implemented by some supermarket managers, benefits of the new regulation should lead to a reduction of food poisoning, which is the cause of millions of dollars in health costs every year and sometimes even death (FSIS 1998b). The new regulation also requires that eggs carry a label stating that refrigeration is required. Therefore, products which are not properly refrigerated during distribution or at the retail level can be charged with misbranding, because a lack of refrigeration would be misleading for customers. In addition, such foods could also be deemed adulterated if held under conditions whereby they may become adulterated, i.e., unfit as food because of the possibility that pathogens could grow under improper storage conditions.

Outdated Products. On May 11-12, 1998, KMSP-TV Minnesota Nine News (UPN) aired an investigative report on shelf-life and freshness dates. The contents of this report included observations of dating practices at 20 local markets of varying sizes. Outdated products were found at 19 of the 20 food stores. On April 28, 1998, refrigerated chicken dated December 1997 was found, and on April 30, 1998, packets of yeast were found that had an expiration date over two years past (April 15, 1996). Kevin Elfering from the Minnesota Department of Agriculture explained that legally, the food just needs to remain wholesome or safe to remain on the shelves even if it is past the expiration date (KMSP-TV 1998). One exception would be eggs, which must be removed by the expiration date, because over time it will not maintain its grade and will be considered misbranded (KMSP-TV 1998). USDA recommends that eggs be dated with a date that is 30 days after laying; this is the time equivalent to one grade loss of the product (FSIS 1998b).

While the dating of eggs is considered a quality and misbranding issue, it would seem reasonable to state that perishable refrigerated chicken found on a grocer's shelves four months past its printed date and yeast (which loses its functional qualities over time) for sale two years past its printed date could be considered as misbranded products. If non-wholesome foods are considered to be misbranded products, then the question is whether or not the open dates on food can be used to define that a product for sale past its recommended open date is beyond wholesomeness, and therefore misbranded.

Omissions of Fact. In GMA v. Department of Public Health (393 NE 2d, 881, 1979), the Supreme Court of Massachusetts ruled that the Massachusetts Department of Health's statutory authority to regulate the sale of food with respect to labels that could be "misleading in any particular" included "labels containing omissions of fact as well as sale of food with labels containing express misstatements of fact." The case was taken up by GMA to overthrow the Massachusetts' requirement for open dates on all foods under the premise of an impediment of commerce between the states. The Court ruled that a food label which is "misleading in any particular" makes the food product misbranded (393 NE 2d 881, 1987). Thus, because there was no date on the package, the food was misbranded, since the purpose of the date was to protect public health.

Therefore, it may be implied under Massachusetts' regulations and in other states with similar legislation on open dating that if a food is not open dated, then it is misbranded since there is an omission of fact important to public health. A food product may also be implied to be misbranded if it is labeled with a date and then not held at proper conditions to meet the promise of that date, i.e., if abused by improper transportation and storage temperatures.

Considering Section 201 (n) of the Food, Drug and Cosmetic Act, if the labeling "fails to reveal facts material in the light of such representations," a product which is purchased in good faith that it is wholesome, but without an open date, may become a non-wholesome product if stock rotation has not been effective or if the distribution temperature is not controlled. If the date does not reveal to the consumer that the product is possibly of a lower quality, then the consumer is not making an intended purchase, because of this omission of information, or misbranding. Such a case has not yet reached the federal court systems, but may be pending in the near future.

CHAPTER 9

PERISHABLE REFRIGERATED PRODUCTS AND HOME PRACTICES SURVEY

Purpose

Funded by The Retail Food Industry Center (TRFIC), a survey was conducted in New Brighton, Minnesota, to collect objective data on consumers' understanding of the following:

- Open dates
- Time-temperature integrators (TTIs)
- Proper home food handling techniques
- Refrigerator rotation techniques

The study was conducted to explore consumer's current level of knowledge regarding open dates, current events in food safety, and home food safety techniques.

Study Design

A two-part, door-to-door survey was conducted in primarily middle-upperclass neighborhoods. The participants were selected randomly because of their proximity of residence to the Silver Lake Cub Foods store, which was using the 3M TTI on its ground beef and prepackaged hamburger products. Most homes in this survey were within two to five miles of the grocery store.

The principal investigator (PI) was accompanied by a second person at all times to ensure her safety while conducting the survey. The second person, who was always female, was to wait outside the participants' homes and to carry the gift certificates and handouts. When the surveyors came to a home where nobody answered the door, they would leave a bright orange handout entitled "Sorry I Missed You..." (Fig. 9.1). If potential participants were busy or wanted additional information before participating in the survey, they were given a bright yellow handout entitled "Perishable Refrigerated Products and Home Practices Survey" (Fig. 9.2).

Part I of the survey was completed by participants in the presence of the PI. Participants received a $5 Cub Foods gift certificate and were given the option of participating in Part II of the survey. Part II required respondents to record their perishable refrigerated food purchases for 2-2½ weeks. As an incentive, a $25 Cub Foods gift certificate was offered upon completion. All names and addresses of participants were recorded and signatures were collected after they received their gift certificates.

Sorry I missed you...
My name is Lynn, and I stopped by earlier today to offer you $5-30 in grocery gift certificates for participating in a quick survey. Why, you ask???

Well, I am a graduate student at the University of Minnesota Department of Food Science and am conducting this survey to learn about consumer food safety and storage practices in the home. If you are interested in more details about the study and would like to participate, please leave a message for me (Lynn) at Labuza's Lab (612) 624-3206. Let me know the best time to call you back, and I'll return your call within 24 hours of your message...

Thank you and have a good day,

Lynn M. Szybist
UMN Graduate Student
Food Science Department

FIG. 9.1. "SORRY I MISSED YOU" HANDOUT

PERISHABLE REFRIGERATED PRODUCTS AND HOME PRACTICES SURVEY

Past studies have indicated that many consumer complaints about food quality may be the result of poor distribution, storage and handling practices at the distribution, retail and home levels. The purpose of this survey is to look into the effectiveness of present efforts at providing consumers "fresher" perishable, refrigerated food products, and to look into the overall food safety knowledge of the participants regarding the consequences of poor rotation and temperature abused conditions.

As a graduate student in the Food Science Department at the University of Minnesota, I will be conducting this study with funds from The Retail Food Industry Center (UMN) and through my advisor at the university, Dr. Ted Labuza. If you have any questions for me regarding this study, please call me at (612) 624-3206. If you want to contact the food science department, you can call the secretary (Gwen) at (612) 624-2792.

FIG. 9.2. "PERISHABLE REFRIGERATED PRODUCTS AND HOME PRACTICES SURVEY" HANDOUT
(Both handouts were reformatted to fit the allotted amount of space.)

Prior to conducting this survey, pilot testing was conducted on a random basis to ensure participants' understanding of the questions and to access the time necessary to complete Part I of the survey. The study design and written materials were then submitted and approved by the University of Minnesota Committee on Human Subjects in Research.

Survey Part I

From June 9, 1999 through July 7, 1999, 101 consumers participated in Part I of the survey. After the PI introduced herself and briefly explained the survey, potential participants were asked to read and sign a consent form before filling in the actual survey. The survey was estimated to take approximately 15 minutes to complete. After participants completed the survey, the PI entered the participants' homes to record the temperatures and models of their refrigerators. The participant was also given the option to record this information if he/she did not feel comfortable inviting the PI into his/her home.

While actually conducting the survey, consumers' responses were often noted and gradually applied as to how to approach prospective participants. For example, in the original script, consumers were immediately informed that they would receive a gift certificate in return for filling-out the survey. While this approach may have attracted student participation, many homeowners immediately thought the surveyors were trying to sell something in return. Most homeowners seemed to be more responsive when the PI introduced the survey in terms of research being conducted at the University of Minnesota and stating lastly that a gift certificate would be given in return for their time. On days when the surveyors wore white, several participants also commented that they initially assumed that they were representing a religious group.

In regard to the time participants actually spent taking the survey, respondents took longer than originally expected. Each survey, including refrigerator temperature measurements, took an average of 25 minutes to complete. The range was about 20-45 minutes per household.

The data from these surveys were initially coded and entered into Excel 4.0. The data were then transferred into Statistical Package for the Social Sciences (SPSS) for statistical analysis.

Sample Population. Demographic characteristics of the sample population were recorded. Compared to the 1994 Continuing Survey of Food Intake by Individuals and 1990 Census data (Table 9.1), which report 51.1% and 51.3% of the U.S. population to be female, 65% of participants in this survey were women. Fifty-five percent of respondents were between the ages of 35-54 years (Fig. 9.3). An impressive 91% of participants had some degree of schooling beyond high school (Fig.9.4) versus the national averages of 46% (CSFII-1994) and 45% (1990 Census) of all U.S. adults. Asking for household income would have been awkward since the PI was entering participants' homes, but estimated figures are provided in Table 9.1. From visual observations, most participants were Caucasian. Almost 89% of respondents often or always did the grocery shopping in their homes (Fig. 9.5).

TABLE 9.1
SUMMARY STATISTICS OF DATA SET (AVERAGE)

	CSFII - 1994	1990 Census
HOUSEHOLD INCOME	$40,440	43,133
Northeast (19.8%)	41,343	47,938
Midwest (23.6%)	41,023	41,597
South (34.6%)	37,738	39,987
West (22.0%)	43,249	45,595
SEX		
Female	51.1%	51.3%
Male	48.9%	48.7%
EDUCATION		
< High School	16.3%	18.3%
High School or GED	25.2%	22.4%
1-3 Years College	16.9%	19.9%
4 Years College	8.5%	9.1%
5 or More Years College	9.3%	4.7%
Child or not asked	23.6%	22.7%

(Carlson *et al.* 1998)

Q: What is your age? (n=101)

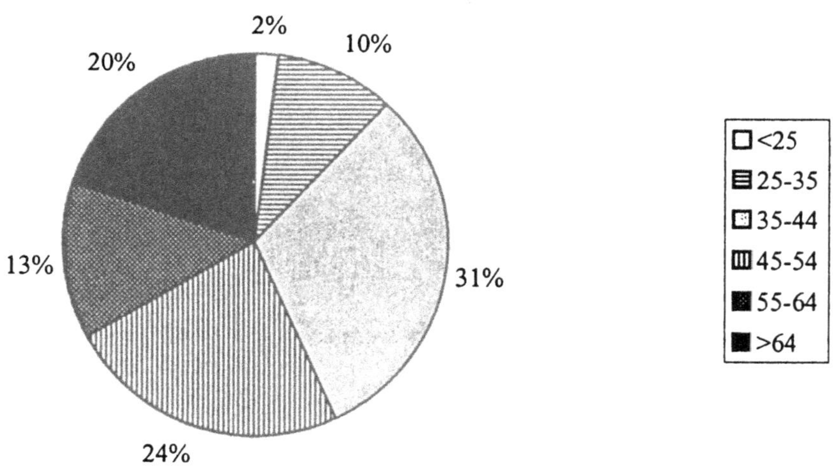

FIG. 9.3. THE SAMPLE POPULATION ACCORDING TO AGE

Q: Which category reflects your education? (n = 101)

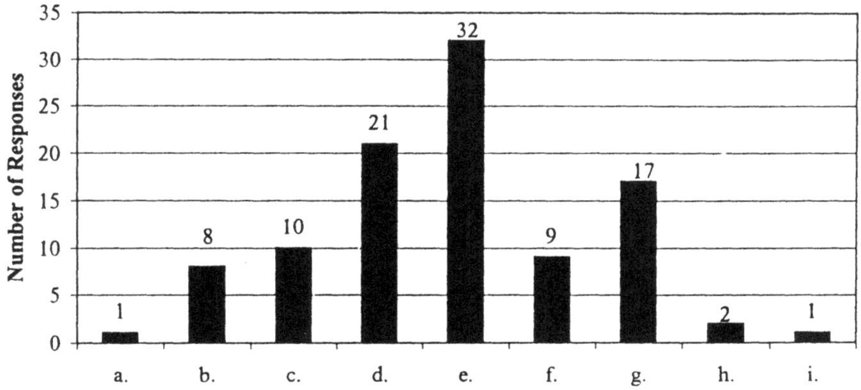

a. <High School Diploma	b. High School Diploma	c. Technical School Degree
d. Some College/Associate's Degree	e. B.S. Degree	f. Some Post Graduate Work
g. M.S. Degree	h. Ph.D Degree	i. Other Advanced Degree

FIG. 9.4. THE SAMPLE POPULATION ACCORDING TO HIGHEST LEVEL OF EDUCATION

Q: How often do you do the grocery shopping for yourself/your household? (n=100)

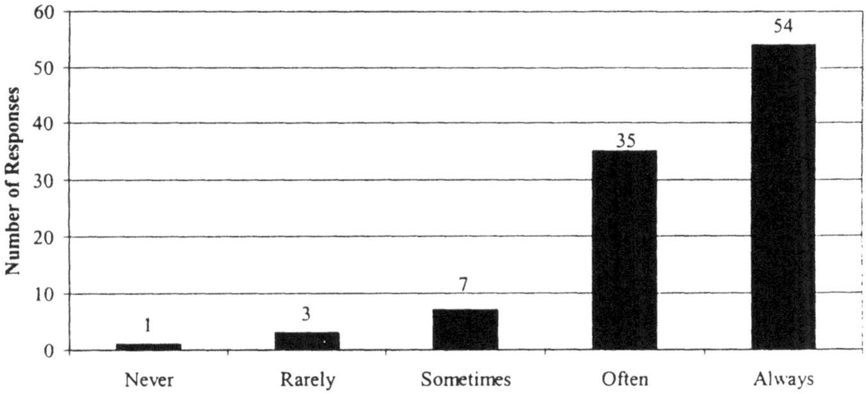

FIG. 9.5. THE SAMPLE POPULATION ACCORDING TO GROCERY SHOPPING FREQUENCY

Results Concerning Basic Food Safety Practices and Knowledge. The first section of the survey focused on participants' practices and knowledge concerning food safety. The results in this survey were compared to results from two extensive food safety surveys conducted in 1997 and 1999 by Audits International (Anon. 1997d; Anon. 1999d). The purpose of the Audits International studies were as follows: (1) to demonstrate the lack of food safety practices in consumers' homes to raise public awareness, (2) to personalize individuals' inadequate home food safety practices, (3) and to encourage more and better home food safety programs.

The 1997 Audits International Survey evaluated the practices of 106 households. Its observations included meal preparation, service, post-meal cleanup and leftover storage practices. The surveyors recorded the participants' violations as major (not likely to cause foodborne illness but often considered as a contributing factor) or critical (can potentially lead to a foodborne illness or injury). Ninety-six percent of the observed households in the 1997 study performed at least one critical violation (Anon. 1997d). Based on Audits International standards, overall less than 1% of households met the minimal criteria for acceptable performance.

To compare the performances in 1997 with the more current study in 1999, the 1997 results were reviewed and gauged against the updated Audits International food safety standards. After modifications were made, households meeting the acceptable performance level in 1997 went from less than 1% to 4%.

In the 1999 survey, the food safety auditors not only looked at the food safety practices of participants, but also attempted to resolve whether the deficiencies in consumers' food safety practices stemmed from a lack of motivation or a lack of knowledge. One hundred and twenty-one households were audited, and unlike the 1997 survey, consumers were asked to answer 4-6 questions regarding food safety. Overall results from the 1999 survey showed that only 69% of households performed a critical violation, and the overall percentage of acceptable performance was 26%. When a critical violation occurred, the follow-up questions revealed that it was usually a result of "lack of knowledge" (62%), rather than from a "perceived lack of importance" (38%) (Anon. 1999d). While the 1999 Audits International Survey demonstrated an increase in consumers' food safety knowledge compared to the 1997 survey, both studies still prove that there is much need for improvement.

Refrigerator Temperature. In this study, the first survey question asked participants what the average temperature of refrigerated foods should be. As shown in Fig. 9.6, about three-fourths of participants knew the correct answer was 40°F. At 20-30°F (which 18 participants chose), foods would freeze, and 50°F (as selected by five individuals) is dangerously high.

Q: What is the average acceptable temperature of refrigerated foods? (n=97)

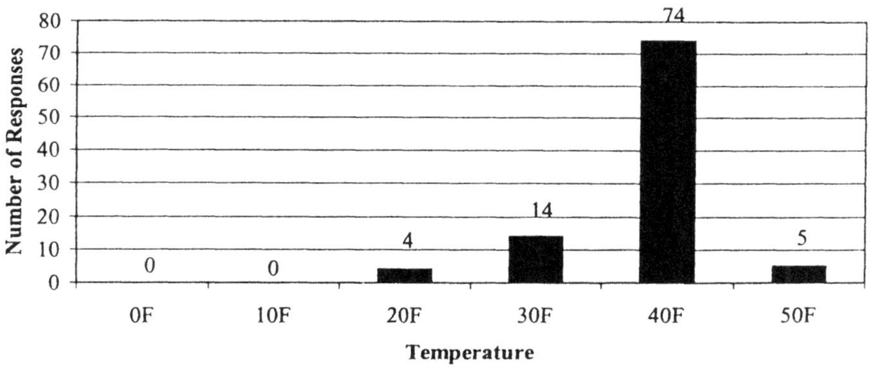

FIG. 9.6. CONSUMER RESPONSES AS TO THE AVERAGE ACCEPTABLE TEMPERATURE OF REFRIGERATED FOODS

In the Audits International Surveys, approximately 65% (1997) and 32% (1999) of participants were cited for violating the recommended maximum refrigerator temperature. Participants were given a major violation if their refrigerators ranged between 42-45°F and a critical violation for temperatures greater than 45°F. Using the same parameters, participants' refrigerator temperatures were recorded in this study using an Atkins Digital Microprocessor Thermocouple Thermometer with Connected Probe (Series 330). Temperature measurements were taken where the milk or an alternative beverage was stored. After placing the thermometer probe under the chosen beverage, the refrigerator door was shut and the temperature reading was taken after approximately two minutes or longer. The measurements ranged from 35°F to 54.9°F with a mean temperature of 41.9°F. The average refrigerator temperature in this survey measured just lower than the 43°F average reported by Audits International in a 1989 survey (Audits International 1990).

Despite the fact that most consumers knew the temperature should be about 40°F, 46% of participants' refrigerators were at an unacceptable temperature (Fig. 9.7). This is better than the 1997 Audits International Survey results but still alarming. Thirty-two percent of participants kept their refrigerators between the unacceptable range of 42-45°F (a major violation), while 14% of refrigerators were dangerously high at above 45°F (a critical violation according to the Audits International Report).

When asked about the significance of proper temperature control, 92% of participants strongly agreed or agreed that maintaining a proper refrigerator temperature was important. Participants felt the most important reason for temperature control was "To *prevent* foodborne diseases" (82%) (Fig. 9.8).

"Keeps foods *longer* (maintain quality)" was the second most important reason (67%), and "*Tastes* better" was least important (81%). Other reasons for maintaining proper refrigerator temperature included preserving the color of the food, preventing food waste, and decreasing the smell or odor of foods.

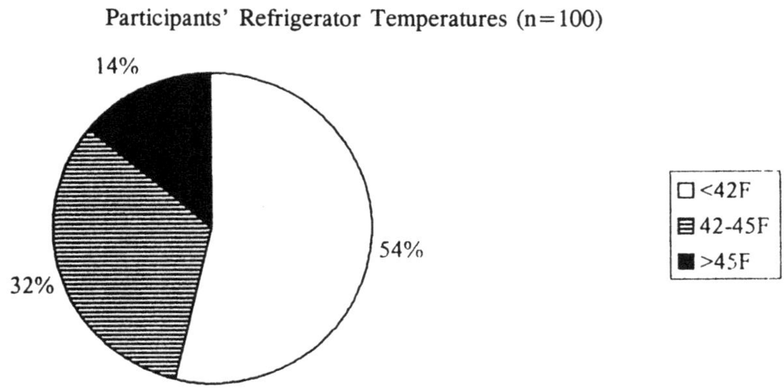

FIG. 9.7. PARTICIPANT REFRIGERATOR TEMPERATURE RESULTS DIVIDED INTO CATEGORIES OF ACCEPTABLE (<42°F), UNACCEPTABLE (42-45°F), AND DANGEROUSLY UNACCEPTABLE (>45°F)

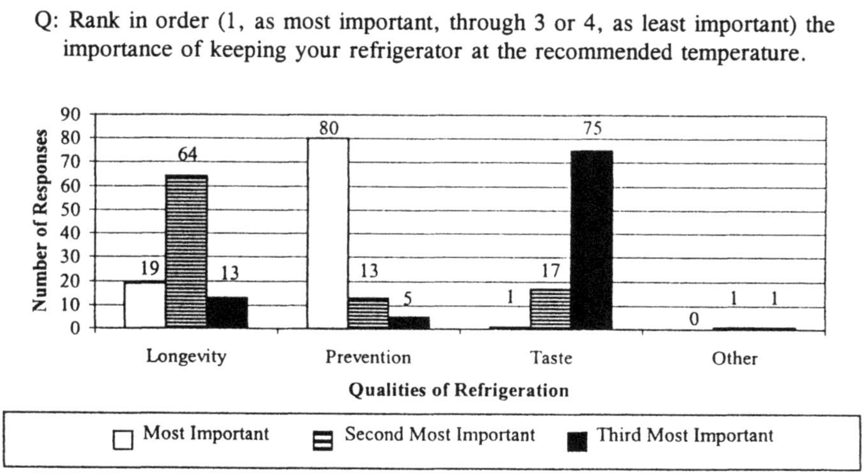

FIG. 9.8. CONSUMER RESPONSES AS TO THE IMPORTANCE OF PROPER REFRIGERATOR TEMPERATURES

In a Nabisco Foods Co. study concerning ingredient and food storage, it was concluded that most homeowners were not aware of the importance of proper refrigeration and the effects of temperature abuse on food (Beard III 1991). In that study, 7 out of 14 homes had refrigerator thermometers, and temperatures ranged from 32-55°F. Only 1 freezer out of 11 had a thermometer, and overall temperatures ranged from 5-20°F. In this study, a sparse 8 out of 101 household refrigerators were equipped with a thermometer. Only 7 out of 97 homes had a thermometer in their freezer; according to these thermometers, freezer temperatures ranged from -5°F to 31.2°F. The average age of refrigerators in this study was 8.9 years, ranging from 3 months to 35 years old. From visual observations of the completed surveys, there was not a significant correlation between the age of refrigerators and proper temperature conditions.

Refrigerated Foods and Recommended Temperature Information on Labels. In an in-store study by Labuza and Szybist (1999a), several refrigerated foods were observed with recommended storage temperatures on their labels. Table 9.2 shows some examples of open dates with temperature information on flavored dip products. The recommended temperatures on Old Home's dip products may be confusing to consumers because in this case, very similar products of the same size were labeled with two different recommended temperature ranges. The printed temperature ranges spanned over 4 and 11°F.

TABLE 9.2
CURRENT OPEN DATING PRACTICES ON FLAVORED DIP CONTAINERS
(Collected on 5/31/98)

FLAVORED DIPS		
Product & Address	***Open Date***	***Printed Dates***
(12 oz) **Gourmet Award Blue Cheese Dip & Dressing** Gourmet Award Foods St. Paul, MN 55114	When properly refrigerated between 33 & 40 this product will retain its wholesomeness for one week beyond date on carton.	6 22
(8 oz) **Old Home's Pride** (Various Flavors) **Dip** Old Home Foods, Inc. St. Paul, MN 55103	Quality assured 7 days beyond date on bottom if properly refrigerated (33-44).	June 2, 1998 to June 30, 1998
(8 oz) **Old Home's Pride** (Various Flavors) **Snack Dip** Old Home Foods, Inc. St. Paul, MN 55103	Quality assured 7 days beyond date on bottom if properly refrigerated (40-44).	June 25, 1998 to July 9, 1998

(Labuza and Szybist 1999a)

When participants in this survey were asked how often they read the labels on refrigerated food products, 59% of participants never or rarely read them, and 25% said they sometimes did. Many people commented that they had never seen a recommended temperature on the label of refrigerated products. When asked if the recommended temperature would affect where they store such products, almost 70% of consumers said that they would stick the product in the refrigerator wherever it fits, regardless if they had read the label or not.

Recent Food Recalls. In the past few years there have been a number of major food recalls that have received widespread media attention. Participants were asked how familiar they were with these events and whether or not these recalls affected their buying habits.

In 1997, Hudson Foods recalled 25 million pounds of ground beef nationwide with suspected *E. coli* O157:H7 contamination (Anon. 1997e). In 1999, 21 deaths (15 adults and 6 miscarriages/stillbirths) were linked to ready-to-eat meat products from Bil Mar Foods, the meat division of the Sara Lee Corp., contaminated with *Listeria monocytogenes* (CDC 1999).

Regarding these meat recalls, 64% of consumers were fairly familiar with the events (Fig. 9.9). According to Fig. 9.10, the recalls affected the buying habits of 53 of the consumers. Of those 53 consumers, 22 respondents stated that they avoided the recalled products for about one to two months, and another 22 participants still did not purchase them. On the other hand, 14% of all respondents stated that they were not afraid of contamination. "Other" comments regarding consumer buying habits spanned from "No, I think it was a bit blown out of proportion" to "I became vegetarian."

In February of 1999, a major recall at a local company, Land O'Lakes, affected Minnesota along with seven other states. On February 10 and 11, the Land O'Lakes Company recalled 10-oz. cartons of milk because of possible *Listeria monocytogenes* contamination, the same bacteria involved in the Bil Mar Foods recall (Anon. 1999e).

The financial losses were minimal compared to the meat recalls, and fortunately there were no reported illnesses or injuries in this incident. Still 58% of respondents were somewhat to extremely familiar with the recall (Fig. 9.11). In this case, only 10% of consumers stated that they avoided the recalled product for one to two months. Despite being somewhat familiar with the recall, 20% of consumers did not avoid buying the milk.

Q: ...How familiar are you with these events (Hudson ground beef and Sara Lee Corp ready-to-eat meat recalls) from the media...? (n=101)

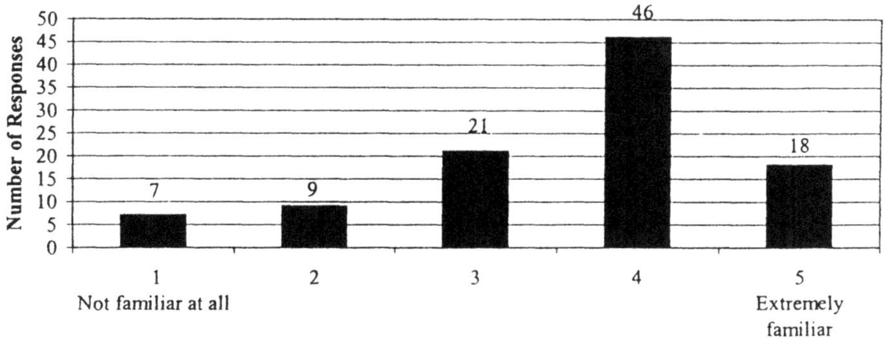

FIG. 9.9. CONSUMER FAMILIARITY WITH RECENT MEAT RECALLS

Q: Did such events affect your buying habits in the meat department? (n=101)

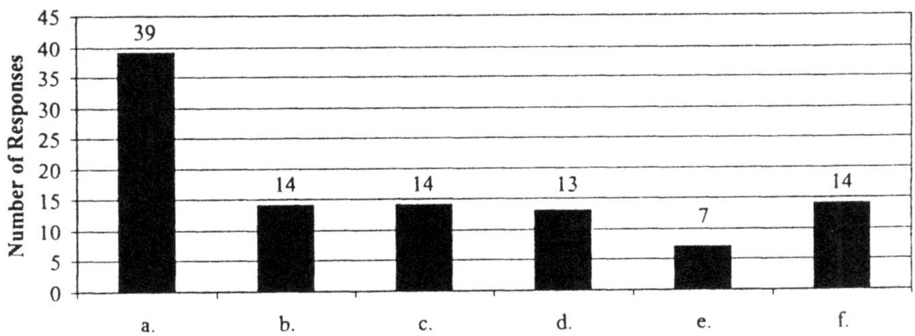

a. Yes, I didn't buy the products from the specific companies involved in the recalls
b. Yes, I didn't buy any of that product regardless of the company
c. No, I still bought the meat because I am not afraid of contamination
d. No, because I was not familiar with the recall
e. No, because I don't buy meat
f. Other

FIG. 9.10. THE EFFECT OF MEAT RECALLS ON CONSUMER BUYING HABITS

Q: How familiar are you with the recent Land O'Lakes milk recall? (n=101)

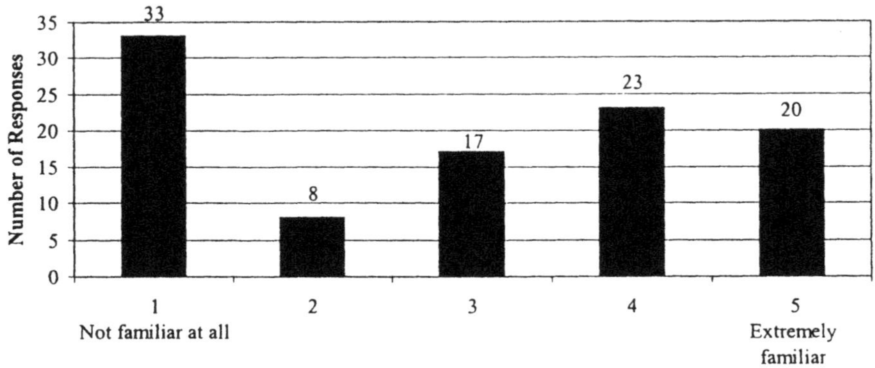

FIG. 9.11. CONSUMER FAMILIARITY WITH THE RECENT LAND O'LAKES MILK RECALL

Meat Preparation. When asked about meat preparation (including beef, pork, chicken and/or fish), 81% of participants prepared meat often (62%), or extremely often (19%). Consumers were then asked about their meat-thawing habits. According to Audits International (Anon. 1997d; Anon. 1999d), there are four ways to properly thaw meats:

(1) In the refrigerator
(2) Under running drinkable water at 70°F or lower within two hours
(3) As part of the cooking process
(4) In a microwave (followed by immediate cooking)

In this study, the meat-thawing practice of "in the refrigerator, the night before use" was the most common method, followed by thawing foods in the microwave. Some participants mentioned that they thaw the meat as part of the cooking process. However, approximately 42% of respondents sometimes or often thawed meat on the countertop the day of use. Such a practice is strongly discouraged, because many food pathogens thrive at room temperature. Consumers also thawed their meat by submerging it in water more often than under cool running water as recommended (Table 9.3).

TABLE 9.3
THE PERCENTAGE OF CONSUMER RESPONSES REGARDING MEAT THAWING PRACTICES
(The most common answer in each row is in boldface.)
Q: How often do you thaw your meat in the following ways? (Check a box in each row, i-v)

	Never	Rarely	Sometimes	Often	Always
i. *In the refrigerator, the night before use* (n=96)	6	18	33	**37**	6
ii. *In the microwave* (n=95)	17	18	**33**	30	3
iii. *On the countertop the day of use* (n=91)	**32**	26	28	14	0
iv. *In the sink submerged in water* (n=92)	**44**	38	14	4	0
v. *In the sink submerged in running water* (n=90)	**61**	24	13	1	0
vi. What other ways have you used?					

Results Concerning TTIs and Ground Beef. As shown in Fig. 9.12, almost 85% of respondents prepared ground beef at least every other week. However, while most participants used sight and smell to test the freshness/safety of ground beef, very few consumers actually tested the internal temperature with a thermometer while cooking (Table 9.4), which is the most effective household way to guarantee its safety. For ground-beef products, the meat must be heated to an internal temperature of 155°F for no less than 15 seconds to ensure that pathogens, which are invisible to the naked eye, are destroyed (Anon. 1997d).

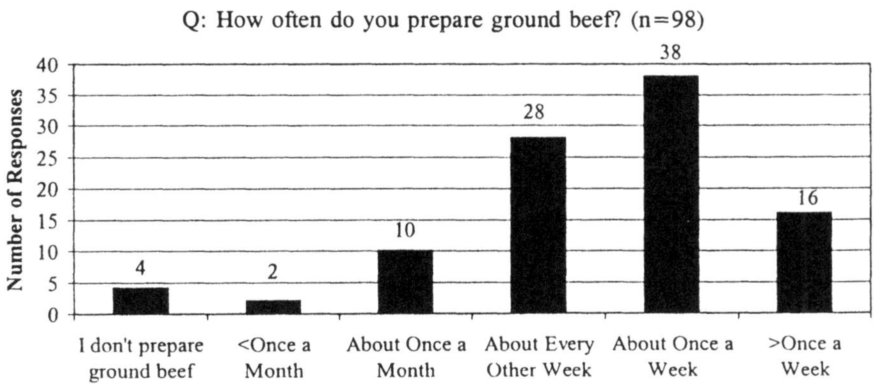

FIG. 9.12. FREQUENCY OF CONSUMER GROUND BEEF PREPARATION

TABLE 9.4
THE PERCENTAGE OF CONSUMER RESPONSES REGARDING PRACTICES USED TO TEST THE SAFETY OR FRESHNESS OF GROUND BEEF
(The most common answer in each row is in boldface.)

Q: How often do YOU test the safety or freshness of your ground beef in the following ways?
(Check a box in each row, i-v)

	Never	Rarely	Sometimes	Often	Always
i. By smelling it (n=95)	6	10	**31**	30	24
ii. By looking at it (n=95)	7	3	23	**37**	30
iii. Use a thermometer to test the internal temperature to 160°F when cooked (n=90)	**77**	11	6	4	2
iv. Rely on the date given on the package (n=94)	14	6	21	**36**	22
v. Don't worry about. Trust that it's safe. (n=85)	**55**	11	19	13	2

Also, while many participants relied on open dates as a measure of safety or freshness, most of them did not trust that the ground beef product was going to be safe when they purchased it (Table 9.4). This is in spite of the fact that 87% of consumers did not purchase any ground beef in the past month where the freshness of the product was questionable (Fig. 9.13).

The "freshness" and safety of ground beef is not only determined by the age of the product, but also by the temperature history to which it is exposed. In November of 1998, the Cub Foods store near the survey location began implementing the 3M time-temperature integrators (TTIs) on some ground beef products. Signs and informative pamphlets were displayed adjacent to meat cases to educate consumers about the tags. The device was also featured in the local newspaper and both local and national television news stories. A picture and a brief description of a TTI were printed directly onto the survey (Fig. 9.14).

Despite the media attention and in-store advertisements, 76% of consumers had never seen TTI's at the supermarket (Fig. 9.15), and the same percentage were not familiar with the device at all (Fig. 9.16). However, after explaining the purpose and function of the tags, all of the consumers believed the device could be somewhat to extremely useful (Fig. 9.17). A few participants were slightly more skeptical when asked if they thought such a device was dependable (Fig. 9.18). Several people commented that the device was probably dependable because it was made by 3M, a local company.

Q: Have you bought any ground beef in the past month where the freshness of the product was questionable? (n=98)

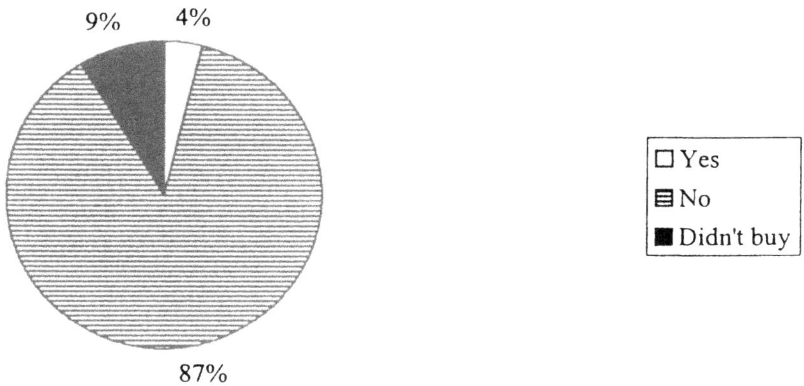

FIG. 9.13. CONSUMER RESPONSES AS TO THE FRESHNESS OF RECENT GROUND-BEEF PURCHASES

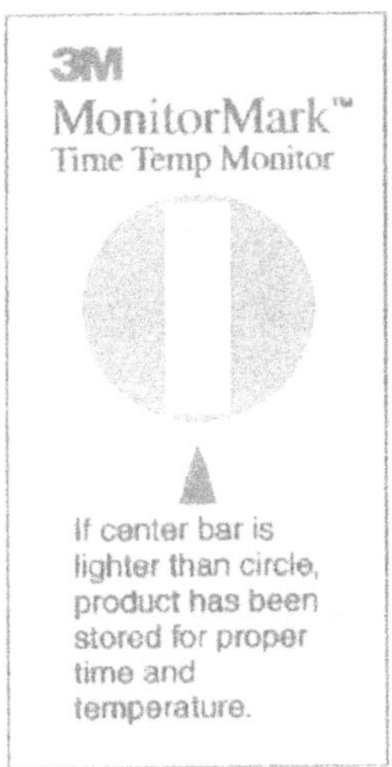

The Fresh Test is a device that looks like a label and is placed on a food package. The "label" changes color over time to indicate abusive (improper) temperature conditions or indicate the end of the product's shelf-life.

FIG. 9.14. PICTURE AND DESCRIPTION OF THE 3M TTI AS PRESENTED TO THE SURVEY PARTICIPANTS

Q: Have you seen the Fresh Test "label" from 3M in the meat department of your supermarket? (n=101)

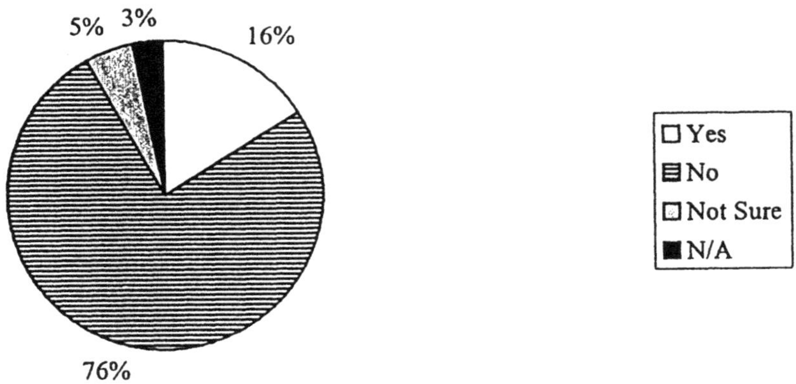

FIG. 9.15. CONSUMER RESPONSES REGARDING WHETHER OR NOT THEY HAD SEEN TTIs IN THE SUPERMARKET

Q: How familiar are you with this "label"? (n=100)

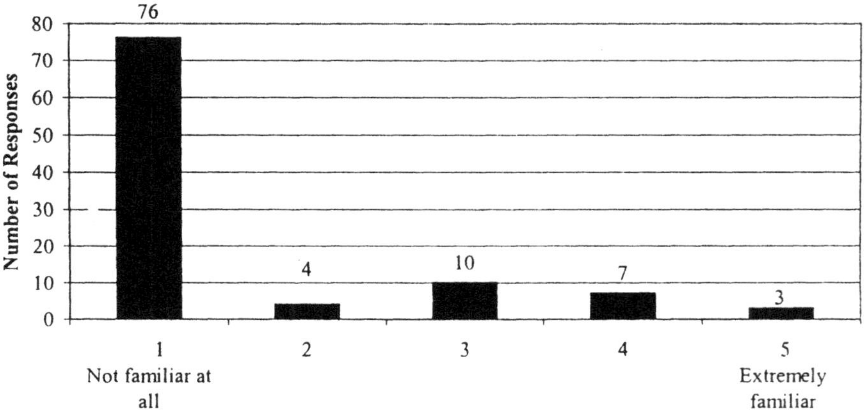

FIG. 9.16. CONSUMER FAMILIARITY OF TTIs

Q: ...Do you believe that such a device can be useful? (n=100)

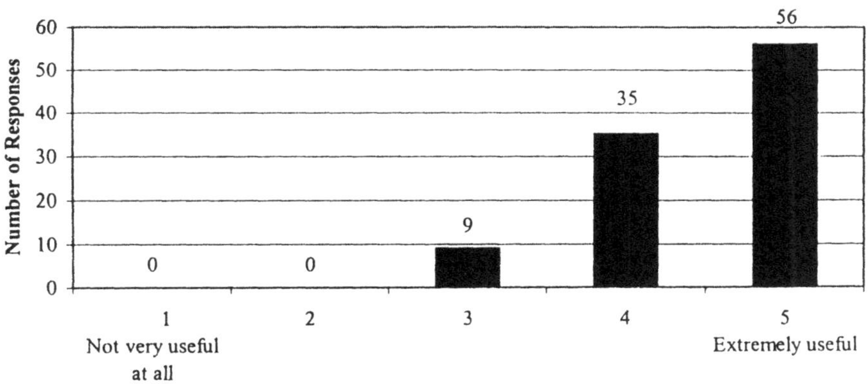

FIG. 9.17. CONSUMER RESPONSES AS TO THE USEFULNESS OF TTIs

Q: Do you think that this device is dependable? (n=99)

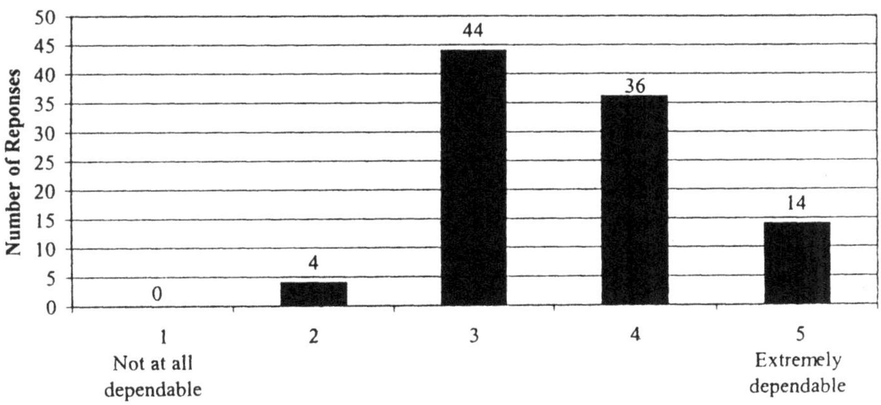

FIG. 9.18. CONSUMER RESPONSES AS TO THE DEPENDABILITY OF TTIs

Results Concerning Open Dates. The final survey questions focused on consumers' use and understanding of open dates. As shown in Fig. 9.19, all participants in this survey claimed to look at open dates on refrigerated products to some extent.

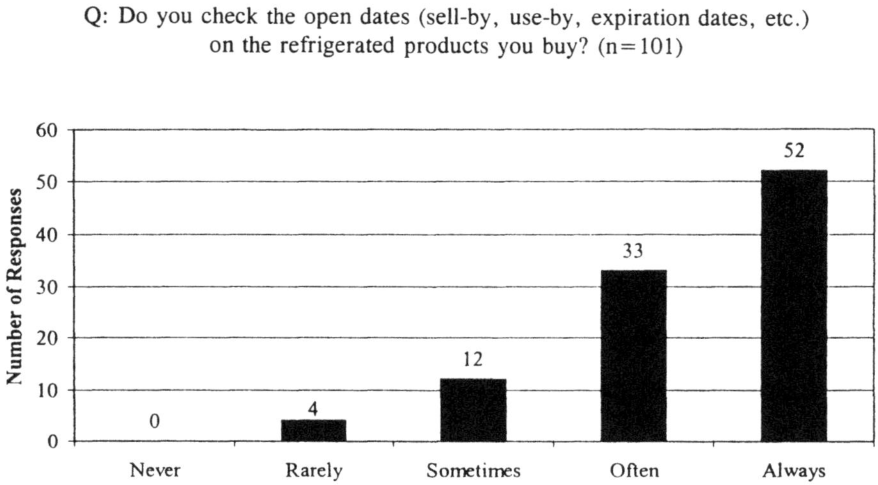

FIG. 9.19. CONSUMER RESPONSES AS TO THEIR FREQUENCY OF CHECKING FOR OPEN DATES

Table 9.5 shows that in this survey, milk was the product that consumers most often checked for an open date. Many respondents stated that on all of the refrigerated products listed, they always checked for an open date.

As stated in Chap. 2, a study in 1971 found that 62% of 628 people sometimes sorted through packages to find the freshest product. The current study found that 81% of participants (n=99) sometimes, often, or always sorted through products to find the package with the longest number of days left according to its given date (Fig. 9.20).

Concerning the reliability of open dates (Fig. 9.21), 65% thought the date was reliable or extremely reliable, and another 33% thought the date was somewhat reliable. This is in spite of the fact that within the past 12 months of this study, at least 36% of participants had purchased refrigerated foods, especially pre-cut salads and vegetables, that had spoiled before the open date (Table 9.6).

TABLE 9.5
THE PERCENTAGE OF CONSUMER RESPONSES REGARDING USE OF OPEN DATES ON PARTICULAR FOOD ITEMS
(The most common answer in each row is in boldface)

Q: How often do you check the open date on the following refrigerated products? (Check a box in each row, i-vi)

	Never	**Rarely**	**Sometimes**	**Often**	**Always**	**I don't buy this product**
i. *Milk* (n=100)	3	4	8	17	**66**	2
ii. *Orange Juice* (n=100)	8	11	14	15	**42**	10
iii. *Ground Beef* (n=100)	5	8	12	16	**51**	8
iv. *Yogurt* (n=99)	2	7	9	17	**44**	20
v. *Pre-cut salads and vegetables* (n=100)	8	12	14	10	**45**	11
vi. *Eggs* (n=99)	13	18	15	10	**40**	3

Q: Do you sort through refrigerated products at the grocery store to find the product with the longest number of days left according to its given date? (n=99)

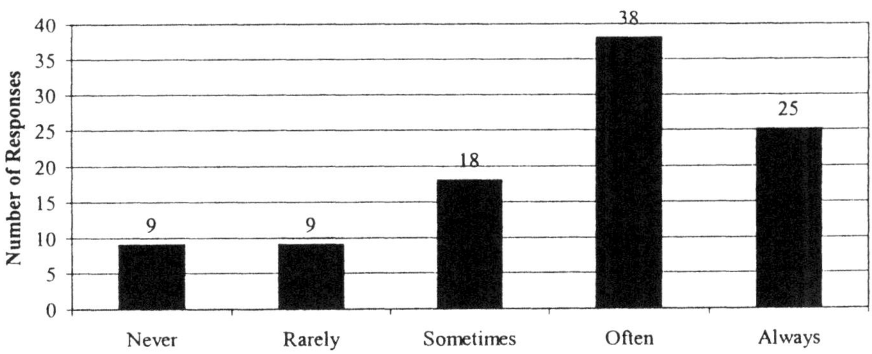

FIG. 9.20. CONSUMER RESPONSES AS TO THEIR FREQUENCY OF SORTING THROUGH OPEN-DATED PRODUCTS

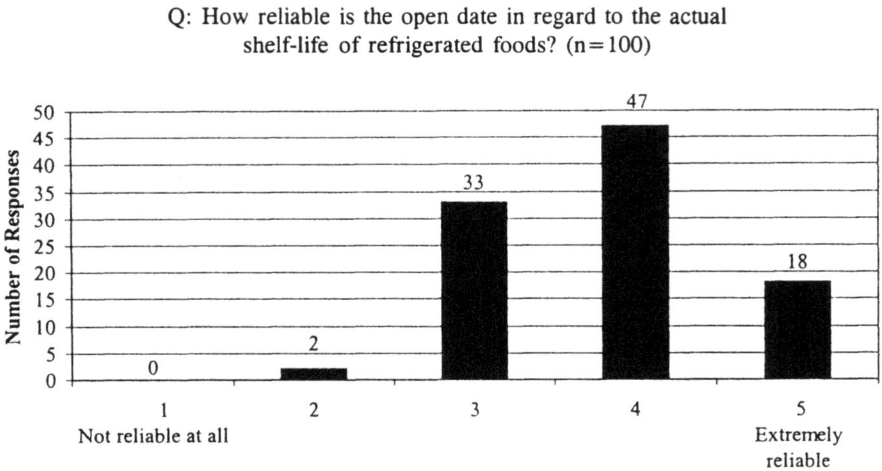

FIG. 9.21. CONSUMER RESPONSES AS TO THE RELIABILITY OF OPEN DATES

TABLE 9.6
THE PERCENTAGE OF CONSUMER RESPONSES REGARDING FOODS SPOILING PRIOR TO THEIR OPEN DATE
(The most common answer in each row is in boldface.)

Q: In the past 12 months, have you purchased any of the following foods and noticed that it spoiled before the open date?
(Check a box in each row, i-vi)

	Yes	No	Didn't notice or consider the date	Didn't purchase
i. *Milk* (n=98)	12	**82**	2	4
ii. *Orange Juice* (n=99)	2	**83**	6	9
iii. *Ground Beef* (n=98)	4	**84**	3	9
iv. *Yogurt* (n=97)	7	**70**	2	20
v. *Pre-cut salads and vegetables* (n=97)	19	**67**	3	11
vi. *Eggs* (n=98)	2	**85**	7	6

The questions in Table 9.7 came from the 1979 OTA study. And just as in 1979, about a third of respondents in this study still did not realize that breakfast cereals were dated. In regard to ground beef, the OTA study claimed that the date referred to the day meat was packaged. However, observations were made at four major grocery stores in the area (Byerly's, Lunds, Cub Foods and Rainbow), and only Byerly's ground beef label printed the day it was packed. The other stores used "sell-by" or "use or freeze-by" dates on their packages. There was also a meat market within New Brighton, MN, which used only the "sell-by" date. The lack of consistency on dating particular food products, such as ground beef, makes it difficult for the consumer to understand the dates. It should be noted that ground beef is not required to be dated under Minnesota law, so such practices were voluntary.

TABLE 9.7
THE PERCENTAGE OF CONSUMER RESPONSES REGARDING THE MEANING OF OPEN DATES ON PARTICULAR FOOD PRODUCTS
(The most common answer in each column is in boldface.)

Q: For each of the following columns (milk, breakfast cereal, and ground beef), please mark the one answer that best represents the date on its package.
(Answer i-iii)

	i. MILK (n=98)		**ii. BREAKFAST CEREAL** (n=98)		**iii. GROUND BEEF** (n=98)	
When it was packaged...	6	(9)	16	(8)	12	**(34)***
Last day it should be sold...	**49**	**(74)***	27	**(35)**	**43**	(31)
Last day it should be used or eaten...	42	(15)	**29**	(26)*	35	(31)
Have never noticed a date on a package of this product...	3	(2)	**29**	(31)	10	(26)

Numbers in parentheses represents 1979 OTA results.
* Asterisks represent "correct" answers

Compared to the OTA study, Table 9.7 shows that a significantly lower percentage of participants understood the meaning of open dates used on milk containers. This was further evident when consumers were asked about their milk storage habits when the product reached the date on the container. Despite the fact that milk is given a "sell-by" date (which represents when the product should be sold at the retail level to give consumers an adequate time for home storage before the end of shelf-life), 16% of consumers poured out their milk on the date printed on the carton regardless. Another 14% threw the milk away the next day. Many respondents (about 19%) relied on unfavorable organoleptic changes to determine the end of shelf-life (Fig. 9.22). "Other" comments included using milk for cooking after the date had passed or that milk never lasted that long in their households.

Q: You buy a container of milk on January 15. The open date printed on the carton is January 20. What do you do with the remainder of the milk on January 20? (n=99)

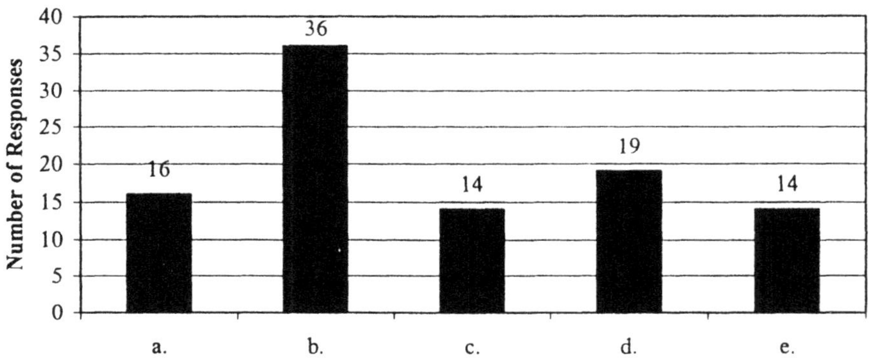

a. Pour out the milk regardless
b. Smell the milk and consume it until it smells bad
c. Throw it out on January 21
d. Continue to drink until it tastes funny
e. __________

FIG. 9.22. CONSUMER RESPONSES AS TO THEIR MILK STORAGE HABITS IN REGARD TO THE OPEN DATE

Survey Part I Discussion and Conclusion. Similar to results reported in the Audits International Surveys, data in this survey concerning participants basic food safety practices and knowledge were insufficient to ensure against foodborne illnesses at home. Temperature abuse was a major source of potential danger as 46% of refrigerators were not at a safe temperature, and 42% of respondents sometimes or always thawed their meat on the countertop.

Furthermore, in regard to ground beef preparation, using a thermometer to test cooked meat products was not a common practice in consumers' homes.

Listeria monocytogenes is another area of concern pertaining to food safety. While a majority of consumers were at least somewhat familiar with the major meat and Land O'Lakes recalls, 20% of respondents did not avoid the Land O'Lakes milk product despite the fact that 21 deaths in the Bil Mar Foods (Sara Lee Corp.) recall were being reported around the same time and involved the same suspected microbial contamination.

Concerning TTIs, although most respondents were not familiar with the device, all participants believed it could be somewhat useful. They also thought that it would be dependable to some extent.

Finally, all respondents looked at the open dates on food labels to some degree, but the misconceptions continued regarding the meaning of the dates. Also, less people understood the open dates on milk containers compared to consumers 20 years ago, despite the fact it is the product consumers most often check for a date. Most consumers believed the date was somewhat to extremely reliable, and 63% of respondents often or always sorted through open-dated products to find food with the longest number of days left until its date. This is in spite of the fact that 36% of respondents had purchased one of the listed food products within the past year which had spoiled before the open date.

Survey Part II

Usually while participants' refrigerator temperatures were being taken for Part I of the study, the PI would explain Part II. Participants interested in continuing onto the second part of the study were then asked to read and sign a consent form. Part II would begin on the day of that participant's next grocery shopping trip. Before putting their groceries away, participants were asked to record their newly purchased refrigerated perishable food items on the sheet(s) provided; directions for properly recording the data were provided on the backside of the recording sheets. Data were to be collected for 2-2½ weeks. Originally, the study required participants to record their food items for 45 days. In the State of Minnesota, open dates must be printed on perishable products if the food's shelf-life is estimated at 90 days or less. Since 90 days seemed like an unreasonable amount of time to expect participant cooperation, the number of days was cut in half. After pilot testing was conducted by two of this study's contributors, including the PI, two weeks was determined to be a reasonable amount of time to gather the necessary data and keep the interest of participants. Since households might not buy groceries and start recording data the very day they were given the survey, the PI returned to collect their data after 2½-3 weeks from the initial visit.

The recording sheets consisted of seven columns. The first five columns asked for the following: (1) "Date of Purchase", the day that the particular food was bought; (2) "(Size) Product Brand & Description", the size of the product as labeled on the package, the product or company name and the common name of the food; (3) "Company Name & Address", which is required by law to be printed on the label; (4) "Explanation of Open Date", meaning the actual description of the date such as "sell-by", "use-by", etc.; and (5) the "Printed Date", or the actual date printed or indented into the package. These columns were to be filled in by participants before they put their perishable groceries into the refrigerator. This included prepackaged fruits and vegetables with a date on them. Any food which was frozen for any period of time, e.g., ground beef, was not to be recorded.

As participants ate or discarded recorded food products, they were to fill in the last two columns. "Your 'end' date" represented the date the food package was discarded. The survey was written this way to avoid confusion. For example, if the consumer used a food product in a recipe, e.g., eggs in cake, the PI was interested in how long it took for the whole carton of eggs to be used and not how long it took to eat eggs in the cake. The final column "Your method of 'End of Product' (e.g., consumed, threw-out, etc.)" asked the participants to record why they discarded the package.

After this explanation, if participants were still interested in participating in Part II, they were handed one or more copies of the recording sheet(s) and given a magnet (Fig. 9.23). The magnet provided the name of the PI and the PIs lab phone number, so participants could call in case they had any questions or concerns.

After approximately 2½-3 weeks, the PI returned to the participating households to collect their recording sheets. The PI was not accompanied by a second surveyor at that time because she did not enter the homes. If the participant was home, the PI collected his/her sheets and asked the participant for a signature in return for a $25 gift certificate. Many homeowners were not present, however, when the PI returned to collect their data. In such cases, the PI put a short message (Fig. 9.24) with a self-addressed and pre-paid envelope in the participants' mailboxes. After receiving their sheets, the $25 gift certificate was sent directly to the households by mail. The data from these surveys were organized and entered into Excel 4.0 for analysis.

Sample Population. The concept for Part II of this survey came from a Nabisco study in 1991 where an inventory study was conducted in 30 households. The goal of this study was to find 50 households to participate in Part II in anticipation that at least 30 households would complete the survey over two weeks. Initially, almost all participants in Part I of this study were asked to participate in the second part. As the number of participants for Part II increased, the PI became more selective and only asked selected participants if they were interested. The selection of participants was generally determined by the participants' level of understanding in Part I and their overall interest in this study.

FIG. 9.23. SURVEY PART II MAGNET

Hello...

I came over today to pick-up part II of my survey on refrigerator rotation practices. Please put your papers in the attached stamped envelope and send them to me at my office at the UMN. As soon as I receive your survey, I'll send you the $25 gift certificate.

Thank you for your participation and your help with my research...

Lynn
UMN Food Science Dept.
(w) (612) 624-3206

FIG. 9.24. LETTER ASKING PARTICIPANTS TO SEND IN PART II
(Document reformatted to fit allotted amount of space.)

A total of 54 of the original 101 participants agreed to partake in Part II of the study. By August 1, 45 participants had completed Part II. Eight of the surveys were unusable, leaving a total of 37 surveys to analyze. From these 37 households, demographic information was collected. As shown in Fig. 9.25 to 9.28, the sample population participating in Part II was representative of the overall sample population in Part I. Both parts averaged 65% female participation with about half of the population between the ages of 35-54 years. Most participants sought some level of schooling after high school and did most of the grocery shopping in their households.

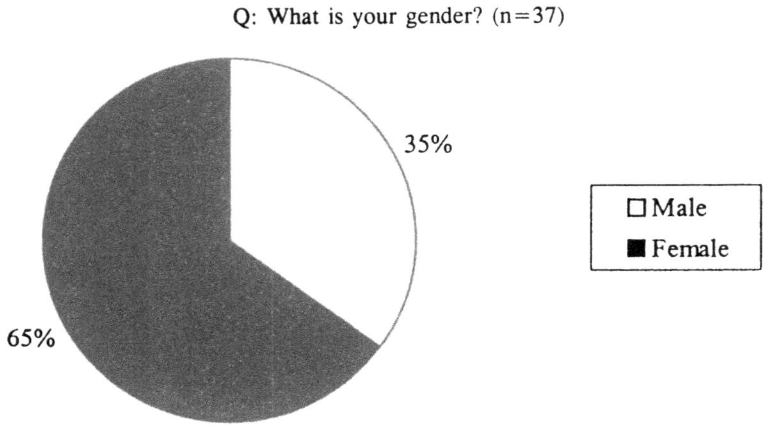

FIG. 9.25. THE SAMPLE POPULATION IN PART II ACCORDING TO GENDER

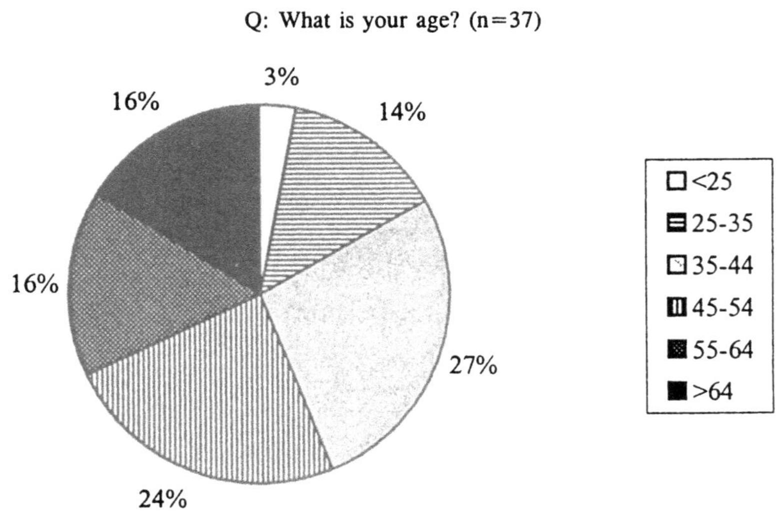

FIG. 9.26. THE SAMPLE POPULATION IN PART II ACCORDING TO AGE

Q: Which category reflects your education? (n=37)

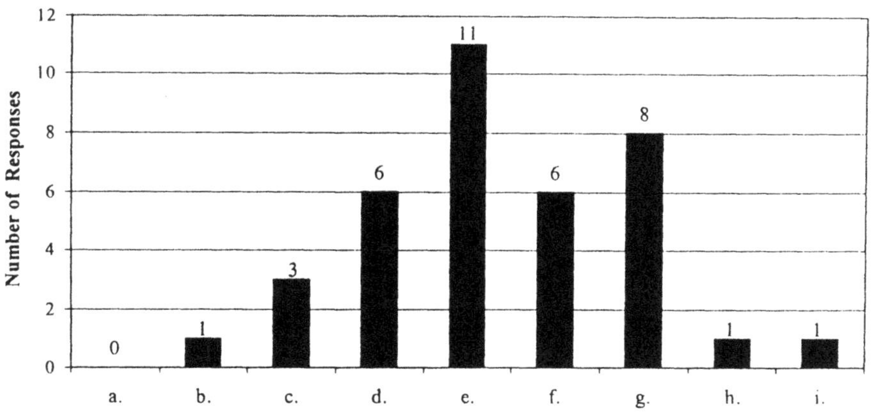

a. <High School Diploma
b. High School Diploma
c. Technical School Degree
d. Some College/Associate's Degree
e. B.S. Degree
f. Some Post Graduate Work
g. M.S. Degree
h. Ph. D Degree
i. Other Advanced Degree

FIG. 9.27. THE SAMPLE POPULATION IN PART II ACCORDING TO HIGHEST LEVEL OF EDUCATION

Q: How often do you do the grocery shopping for yourself/your household? (n=37)

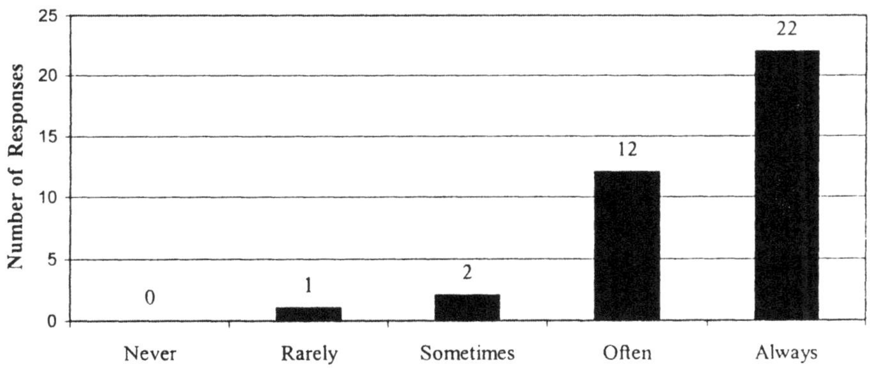

FIG. 9.28. THE SAMPLE POPULATION IN PART II ACCORDING TO GROCERY SHOPPING FREQUENCY

Storage Results.

Pantry Closets Rotation Study. The purpose of Part II of this study was to better understand the rotation habits of consumers regarding perishable refrigerated foods. The idea was inspired by a study published by T.D. Beard III at Nabisco Foods Co. in 1991. His overall study was conducted to better determine the reason for increased complaints and customer dissatisfaction. As part of his study, Beard III evaluated the home storage practices of 30 households and their pantry closets. The study revealed "home warehouse" as the least understood area of the kitchen (Beard III 1991). Table 9.8 shows results from his study.

TABLE 9.8
STORAGE TIME OF PRODUCTS IN CONSUMER PANTRIES

PRODUCTS	STORAGE TIME (WEEKS)	
	Average	Range
Canned Goods	12.3	1-104
Ethnic Foods	10.9	1-52
Condiments	11.6	1-156
Baking Products	21.6	1-260
Dry Goods/Pasta, Mixes	17.7	1-156
Breakfast Cereals	12.0	1-150
Cookies, Crackers, Snacks	7.6	0.3-26
Coffee, Juices, Beverages	36.3	1-156
Dog/Cat Food	11.0	1-28
Paper Products	--	1-52

(Beard III 1991)

Careless storage practices were a common problem, especially with baking products, dry goods, condiments and breakfast cereals. This may lead to increased consumer dissatisfaction of products, stale and rancid products, and the presence of insects. Rotation of food products was extremely poor and many households did not date their products. An open date may not only assist retailers with stock rotation, but may also be beneficial in managing household stocks (Beard III 1991).

Beard's study examined non-refrigerated products, but the importance of open dates and stock rotation practices of perishable, refrigerated products is also extremely important because of their limited shelf-lives.

Perishable Refrigerator Products Rotation Study. Data from participants' recording sheets were reviewed and organized by product category. Milk, orange juice, ground beef, yogurt, pre-cut salads and vegetables, and eggs were common products recorded by the respondents and were further analyzed. Cheese was also a common purchase, but the shelf-life of the product was generally longer than the extent of this study.

Table 9.9 shows the results from Part II. The first two columns provide information similar to the Beard study (Table 9.8). The third column represents the number of days between the date of purchase and the open date on the product. In the final column, the number of days between the actual 'end' date of the product and the open date were calculated. Note that the figures used to calculate the "average" number of days and the "ranges" were limited to those products eaten or discarded during the 14-18 days of this study.

Milk. Milk was the most popular purchase recorded. Out of 108 total samples, 85 cartons reached the 'end' point during the duration of this study. Three of the cartons were discarded 1-2 days after the printed date. One consumer noted that a container of milk was discarded eight days *before* the printed date because it smelled funny. Another participant consumed the milk eight days *after* the printed date.

The average home storage was 6.3 days. In one household, the milk was stored for three weeks before it was completely consumed. Most consumers purchased the milk with about 13.3 days remaining before the printed date on the carton.

By the time most households reached the end of the product, there were 7.1 days remaining until the printed date. For the most part, the containers were ½ gallon to one gallon in size, and a "sell-by" date was usually printed on the milk cartons, (there was one "use-by" date recorded and a few dates without an explanation). In other words, most milk products in this study had not even reached the "sell-by" date or the date when the product should be removed from the grocers' shelves to still have an adequate amount of time for home storage and consumption. For the three products that were discarded 1-2 days after the date, the product was probably temperature abused or thrown-out because of a misunderstanding about the date's actual meaning. The products were also sitting in the refrigerator for 10, 10, and 15 days since the day of purchase. If the milk container was opened on one of the first days it was purchased, then it is possible that the milk did spoil. As for the milk that was discarded eight days before the "sell-by" date, the product had also been in the consumer's refrigerator for 7 days. If the container was opened for several days and the refrigerator temperature was too high, this could explain the premature spoilage.

TABLE 9.9
STORAGE TIME OF PRODUCTS IN CONSUMER REFRIGERATORS

Product	**Home Refrigeration Storage Time (Days)**		**Time between Purchase Date & Open Date (Days)**		**Time between 'End' Date & Open Date (Days)**	
	Ave.	*Range*	*Ave.*	*Range*	*Ave.*	*Range*
Milk (n=108)	6.3	1-21	13.3	3-18	7.1	(-8)- 15
Orange Juice (n=16)	8.9	4-14	37.3	6-55	28.3	(-3)- 43
Ground Beef (n=12) *(n=7)**	4.8 *(0.5)**	1-22 *(0-1)**	0.8 *(0.9)**	0-2 *(0-2)**	-4.0 *(0.3)**	(-21)- 0 *((-1)-1)**
Yogurt (n=43)	5.4	0-19	21.1	3-34	15.4	(-8)- 34
Pre-cut	5.6	0-17	7.0	3-13	1.0	(-5)- 7
Eggs (n=12)	8.7	1-17	22.2	9-36	12.6	2-22

Negative numbers represent the number of days past the printed date
* The numbers in parentheses were averaged after five responses were removed from the calculations

Orange Juice. From the 16 recorded containers of orange juice, 11 of them were consumed and none were discarded. The orange juice was stored for an average of 8.9 days in the refrigerator. When most consumers purchased their juice, there was an average of 37.3 days left before the product even reached its printed date. There was only one "use-by" date written and three "Best if sold by" dates. The rest of the open dates were "sell-by". Even by the time the consumer finished the product, there was still an average of 28.3 days remaining before the open date. Only one consumer indicated that the product was consumed after the "sell-by" date.

The open shelf-life dates on orange juice are significantly longer than the dates on milk. Many of the orange juice samples in this study had over a month left before their printed dates. Therefore, even if some products suffered from slightly temperature-abused conditions (where the product would experience a shorter shelf-life), all of the juice recorded in this study was consumed before any negative sensory changes were noticed.

Ground Beef. There were two averages calculated for the ground beef data. The top set of data represents figures from all 12 recorded samples, while the second set of numbers in parentheses are from only seven of the samples. After reviewing all of the data, it was assumed that some of the ground beef was frozen and erroneously recorded. All of the figures for the ground beef data were from the "sell-by" date, and not from the "use or freeze by" date that was also printed on some packages.

From the 12 samples in this study, the average number of refrigerator storage days was 4.8. The range of refrigerator storage, however, was from one to a questionable 22 days. In fact, five of the samples supposedly were stored in the refrigerator for seven days or more. Most products were purchased with only 0-2 days until the open date, and product was consumed up to 21 days after the open date.

The figures in parentheses represent the seven products with less than seven days of refrigerator storage. From these figures, the ground beef averaged only 0.5 days of refrigerator storage. The actual 'end' date of the product ranged from "one day before" to "one day after" the "sell-by" date, and overall these products were consumed an average of 0.3 days before the open date.

Yogurt. From the 57 recorded samples of yogurt, 43 were consumed by the end of this survey and only one was discarded. Most yogurt was stored in the refrigerator for about 5.4 days, and had an average of 15.4 days until reaching the "sell-by" date. Most products contained a "sell-by" date or only a printed date, e.g., the Dannon brand.

Only three people consumed products after the "sell-by" dates, indicating there were sufficient rotation practices at the retail and home levels to move the product quickly. In the only sample that was discarded, there were still 16 days until the printed date on the container, but the consumer threw-out some of the product because it "didn't look good." The product was a 32-oz container of yogurt, so most likely, the container was opened several times before finally reaching an unacceptable appearance. There were 12 yogurt samples that were 28 oz or larger.

Pre-Cut Salads and Vegetables. Packaged mushrooms and baby carrots were recorded in this section, but the mushrooms contained a "packed on" date (which would skew the data) and the shelf-life for carrots was significantly longer than the other recorded products; therefore, the open dates on these products were not evaluated.

The products in this category were all most likely CAP/MAP to extend shelf-lives. On average, the products were purchased with just a week left to the open date. "Use-by" or "best-if-used-by" dates were most commonly used.

Out of 20 recorded samples, 13 were fully consumed and one was discarded because of browning. On average, there was only one day left from the open date to the actual date of final consumption. One product was consumed five days after the "best-if-used-by" date.

Eggs. Out of 21 samples, only 12 cartons of eggs were consumed during the allotted period of time. From the consumed samples, the average number of refrigerator storage was 8.7 days. That figure is probably lower than the actual average because many participants were still consuming their eggs after the 2½ week recording period. All of the recorded figures showed that the eggs were consumed before the expiration date. A few of the packages contained "sell-by" dates.

Part II Conclusion and Discussion. Out of 204 perishable refrigerated food products purchased, recorded, and evaluated in Part II of the survey, there was a total of six products discarded. Two products were discarded before the open dates, three were discarded one or two days after the "sell-by" dates, and one was on the "best-if-used-by" date. Although it was not possible to determine the exact reasons why these products had to be wasted, temperature abuse may have been a contributing factor. Four of the discarded products were milk, one was a pre-cut salad, and the other was a container of yogurt. Whether these products were abused in the summer heat during transportation or stored at improper refrigerator temperatures, according to the open date, none of these products should have been spoiled if they were kept at proper temperature conditions.

Overall, most foods were purchased with a significant number of days left until the printed date. This was especially true for the sell-by dates on most orange juice and yogurt products as well as the expiration dates on most of the eggs. All of the products were purchased before the printed date, which indicates that either there were efficient stock rotation practices at the grocery store level or that the consumers sorted for the youngest products.

CHAPTER 10

CONCLUSION

Since 1979, some states have introduced mandatory open-dating legislation, suggesting a growing awareness of its potential benefits, an awareness consumers have been conscious of for almost a century. However, the lack of uniformity among state laws leads to confusion as to the meaning and proper usage of open dates. Consumers' lack of understanding regarding open dates was evident in our recent survey (Labuza *et al.* 2001), in which all participants responded that they have looked at a food package to find the open date, however, many of the respondents did not use the dates properly. The lack of consistency among dating practices within the food industry has strongly contributed to the overall misunderstanding of food dating, and until federal dating standards are imposed, it is only a matter of time before the inaccuracy or omission of an open date leads to a complicated lawsuit against a food company or grocery chain.

The benefits behind mandatory open-dating legislation, such as we propose, make it the ideal system to move retailers into the next millennium. As the supermarket industry strives towards more efficiency in their distribution and internal structure, the implementation of open dating would complement the goal of ECR. At the same time, the technology of TTIs used in conjunction with an open date would increase the validity of a wholesomeness guarantee for high-quality food products. The inconvenience for manufacturers in adding a uniform open-dating system to their products would be minimal compared to the benefits of educating consumers and building consumer trust not only toward the supermarket industry, but with respect to the freshness of this country's food.

REFERENCES

A.C. Nielsen Co. 1973. Study of consumer attitudes toward product quality. Northbrook, IL.

Anon. 1971. Food stability and open dating. Food Science Dept., Rutgers Univ., New Brunswick, N.J.

Anon. 1979. Open dating and food waste. Minneapolis Tribune, Oct.

Anon. 1990. Time temperature devices: USDA requirement urged. Food Chemical News, Oct. 15, p. 36.

Anon. 1997a. Prevention Magazine — NBC Today Survey. Prevention Magazine 49(8): 86–95.

Anon. 1997b. Born-on dating: A fresh strategy. Food Processing, Feb., p. 29.

Anon. 1997c. Uniform Open Dating Regulation. National Institute of Standards and Technology. Handbook 130.

Anon. 1997d. Audits International's Home Food Safety Survey. (http://www.audits.com/survey.htm).

Anon. 1997e. Hudson recall reflects system's cracks in record-keeping, rework under question. The Omaha World-Herald Company. December 16.

Anon. 1998a. How $100 is spent in the supermarket. Food Processing, Nov., p. 12.

Anon. 1998b. Retail temperature control lags food code. Food Reg. Weekly 1(2): 3–4.

Anon. 1999a. Bad politics, bad meat. Los Angeles Times, Feb. 12.

Anon. 1999b. National Enquirer. Aug.

Anon. 1999c. LifeLines: Frequently asked questions. http://www.lifelinestechnology.com/faq.html.

Anon. 1999d. Audits International's Home Food Safety Survey. (http://www.audits.com/Report.html).

Anon. 1999e. Land O'Lakes recalls milk in eight states. Star Tribune (Minneapolis, MN). Feb. 4.

Anon. 2000. Alabama Safe Foods Act of 2000. (http://www.alabamagrocers.org/safefoodacts.htm).

Anon. 2001. Food Safety News (FS Net), June 12, University of Guelph, Guelph, Ontario, Canada.

Audits International. 1990. 1989 National retail food product cold temperature evaluation. Highland Park, IL.

Beard III, T.D. 1991. HACCP and the home: The need for consumer education. Food Technol. 45(6): 123.

Bernard, D.T. 1997. Hazard analysis and critical control point system — Use in controlling microbiological hazards. *In* Food Microbiology Fundamentals and Frontiers, Chap. 41, ed. M.P. Doyle, p. 740–741. American Society for Microbiology, Washington, D.C.

Besfamille, M. 1998. Personal communication. Univ. of Toulouse, France.

Bishai, D.M., Bhatt, S., Miller, L.T. and Hayden, G. 1992. Vaccine storage practices in offices of pediatrics. Pediatrics 88: 193–196.

Brody, A.L. 1997. Chilled foods distribution needs improvement. Food Technol. 51(10): 120.

Brody, A.L. 1998. Minimally processed foods demand maximum research and education. Food Technol. 52(5): 66.

Butler, J. 1998. Sainsbury's fined for selling stale food. PA News, June 23, p.2.

Campden. 1992. A food industry specification for defining technical standards and procedures for evaluation of time temperature indicators. Tech. Manual 35. Campden Food and Drink Assn., Campden, U.K.

Carlson, A., Kinsey, J. and Nadav, C. 1998. Who eats what, when, and from where. Working Paper 98-05. Sloan Foundation, The Retail Food Industry Center, Univ. of Minnesota, St. Paul.

CDC. 1999. Update: Multistate outbreak of listeriosis. Centers for Disease Control and Prevention (www.cdc.gov/od/oc/media/pressrel/r990114.htm).

Cherng, Y.S. and Zall, R.R. 1989. Use of time-temperature integrators to monitor fluid milk movement in commercial practice. Dairy, Food and Environmental Sanitation, Sept., p. 439.

Cox, J. 1998. U.S. armed services consider food safety; leading-edge technology is featured. (Press Release) Boston, MA, Oct. 26.

Dern, A. 1998. California's Proposition 65 would feel the heat of national uniformity for food bill. Food Chemical News, Aug. 10, p. 22–3.

Dowdell, S. 1996. Looking for a date. Supermarket News, June 17, p. 27–36.

Duyvesteyn, W. 1997. Integration of time-temperature history effect on the shelf life of fluid milk. M.S. Thesis, Univ. of Minnesota, St. Paul.

FDA. 1979. Summary of hearings on open dating. Food and Drug Admin., Fed. Reg. 44: 76007.

FDA. 1994. Fish and fishery products hazards and controls guide. Food and Drug Admin.,Washington, DC.

FDA. 1995. Food Code. Food and Drug Admin., U.S. Department of Commerce, Springfield, VA.

FDA. 1996. Transportation and storage requirements for potentially hazardous food. Food and Drug Admin., Fed. Reg. 61: 59372–59382.

FDA. 1998. Guide to minimize microbial food safety hazards for fresh fruits and vegetables (Draft Guidance). Food and Drug Admin., Washington, DC.

Find/SVP. 1988. Time-temperature monitoring products. A comparative intelligence report. Find/SVP, New York.

FNQUEB. 1999. Energy efficiency report. Far North Queensland Electricity Corp., Ltd. (www.fnqeb.com.quindex.html).

FSIS. 1993. Mandatory safe handling statements on labeling of raw meat and poultry. Food Safety and Inspection Service, U.S. Dept. of Agriculture, Fed. Reg. 58: 43478–43489.

FSIS. 1995a. Focus on food product dating. Food Safety and Inspection Service, U.S. Dept. of Agriculture, Washington, DC. (www.fsis.usda.gov.80/oa/pubs/dating.htm).

FSIS. 1995b. Use of the term "fresh" on the labeling of raw poultry products. Food Safety and Inspection Service, Fed. Reg. 60: 3454–3462.

FSIS. 1996. Pathogen reduction; hazard analysis and critical control point (HACCP) systems. Food Safety and Inspection Service, U.S. Dept. of Agriculture, Fed. Reg. 61: 38805–38855.

FSIS. 1998a. Guidance for beef grinders to better protect public health. Food Safety and Inspection Service, U.S. Dept. of Agriculture, Washington, DC. (www.usda.gov/fsis/oa/haccp/guideb.pdf).

FSIS. 1998b. Refrigeration and Labeling Requirements for Shell Eggs. Food and Drug Admin., Fed. Reg. 63: 45663–45675.

Hartman, P.A. 1997. The evolution of food microbiology. *In* Food Microbiology Fundamentals and Frontiers, Chap. 1, ed. M.P. Doyle, p. 10. American Society for Microbiology, Washington, DC.

IFT. 1981. Open shelf-life dating of food. Inst. of Food Technologists, Food Technol. 35(2): 89–96.

IFT. 1998. Hark! The herald food microbiologists sing! Inst. of Food Technologists, IFT News Release. December 15.

Johnson, A.E., Donkin, A.J.M., Morgan, K., Lilley, J.M., Neale, R.J., Page, R.M. and Silburn. R. 1998. Food safety knowledge and practice among elderly people living at home. J. Epidemiol. Community Health, p. 745–748.

Kalish, F. 1991. Extending the HACCP concept to product distribution. Food Technol. 45(6): 119.

Kantor, L.S., Lipton, K., Manchester, A. and Oliveira, V. 1997. Estimating and addressing America's food losses. Food Rev. 20 (Jan.-Apr.): 2–12.

KMSP-TV. 1998. Food shelf-life. Minnesota Nine News. May 11–12.

Krassie, J.G. 1999. Food temperature audits in five Australian institutions. Foodservice Research International 11(2): 95–105.

Kurt Salmon Associates. 1993. Efficient Consumer Response (enhancing consumer value in the grocery industry). Presented at Food Marketing Institute, Washington D.C.

Labuza, T.P. 1982. Shelf-Life Dating of Foods. Food & Nutrition Press, Trumbull, CT.

Labuza, T.P. and Schmidl, M.K. 1985. Accelerated shelf-life testing of foods, Food Technol. *39*(9) 57–62, 64, 134.

Labuza, T.P. and Schmidl, M.K. 1988. Use of sensory data in the shelf-life testing of foods: Principles and graphical methods for evaluation. Cereal Foods World 33(2): 193–205.

Labuza, T.P. and Szybist, L.M. 1999a. Current practices and regulations regarding open dating of food products. Working paper 99–01. Sloan Foundation, The Retail Food Industry Center, Univ. of Minnesota, St. Paul.

Labuza, T.P. and Szybist, L.M. 1999b. Playing the open dating game. Food Technol. 53(7): 70–85.

Labuza, T.P., Szybist, L. and Peek, J. 2001. Perishable refrigerated products and home practices survey. Working Paper 01-04. The Retail Food Industry Center, Univ. of Minnesota, St. Paul.

Leek, F.W. and Ronnow, P. 1999. Ground beef studies/time-temperature indicators (preliminary report). http://www.vitsab.com/VITgroundbeef.htm.

Malcata, F.X. 1990. The effect of internal temperature gradients on the reliability of surface mounted full history time-temperature indicators. J. Food Proc. Preserv. 14: 481–487.

McWilliams, M. 1993. Foods — Experimental Perspectives, 2nd Ed. Macmillan Publishing Company, New York.

Mistry, V.V. and Kosikowski, F.V. 1983. Use of time temperature indicators as quality control devices for market milk. J. Food Protect. 46(1): 52–57.

Moore, A.K. 1997. Dated thinking. Supermarket News, Mar. 24, p. 40.

Narisetti, R. 1997. P&G's new no-returns policy tells retailers to keep damaged goods. Wall St. J., Mar. 24.

NIST. 1997. Institute of Standards and Technology. (http:www.nist.gov/).

NSFI. 1997. Food safety systems without borders. The Standard of Excellence (3rd Q). National Sanitation Foundation International, p. 4.

OTA. 1979. Open shelf-life dating of food. Office of Technology Assessment (OTA). U.S. Govt. Print. Office, Washington, D.C.

Pryce, R. 1987. The dynamics of European Union. Croom Helm, New York.

Seligsohn, M. 1979. Smashing the open-dating myth. Food Eng. 10: 20–25.

Sherlock, M., Fu, B., Taoukis, P.S. and Labuza, T.P. 1991. A systematic evaluation of time-temperature integrators for use as consumer tags. J. Food Protect. 54(11): 885–889.

Sherlock, M. and Labuza, T.P. 1992. Consumer perceptions of consumer type time-temperature indicators for use on refrigerated dairy foods. Dairy, Food Environ. Sanitation 12: 559–565.

Silverman, J. 1999. Interview with Tom Billy. Food Safety Report ISSN 1523-4533, Bureau of National Affairs, Washington, D.C., Feb. 17.

Skinner, G.E. and Larkin, J.W. 1998. Conservative prediction of time to Clostridium botulinum toxin formation for use with time-temperature indicators to ensure the safety of foods. J. Food Protect. 61: 1154–1160.

Springen, K. 1998. Safer food for a tastier millennium. Newsweek, Sept. 28, p. 14.

Stamp, J.A. 1990. Kinetics and analysis of aspartame decomposition mechanisms in aqueous solutions using multiresponse methods. Ph.D. Thesis, Univ. of Minnesota, St. Paul.

Stickel, A.I. 1996. Values that make cents. Supermarket News, July 1, pp. 15, 18–19.

Taoukis, P.S., Fu, B. and Labuza, T.P. 1991. Time-temperature indicators. Food Technol. 45(10): 70–82.

Taoukis, P.S. and Labuza, T.P. 1989a. Applicability of time temperature indicators as shelf life monitors of food products. J. Food Sci. 54: 783–788.

Taoukis, P.S. and Labuza, T.P. 1989b. Reliability of time temperature indicators as food quality monitors under non-isothermal conditions. J. Food Sci. 54: 789–793.

Taoukis, P.S. and Labuza, T.P. 1996. An integrated approach to food chemistry. *In* Food Chemistry, ed. O. Fennema, p. 1013–1042. Marcel Dekker, New York.

Taoukis, P., Labuza, T.P. and Saguy, I.S. 1997. Kinetics of food deterioration and shelf-life prediction. *In* Handbook of Food Engineering Practice, editors K. Valentas, E. Rotstein and R.P. Singh, CRC Press, Boca Raton, FL.

USDA. 1973. Food dating: Shoppers' reactions and impact on retail food stores. Marketing Research Report. Econ. Res. Service, U.S. Dept. of Agriculture, Washington, DC.

USDA. 1996. Food standards and labeling policy book. Aug.

Williams, M. 1998. Simple advice. Supermarket News, July 13, p. 31–32.

APPENDIX A

Proposed 1999 TTI Legislation in Italy

Bill Brought Before Italian Parliament by Some Senators:

(Bucciero, Caruso A., Maceratini, Basini, Battaglia, Bevilacqua, Bonatesta, Bornacin, Bosello, Campus, Castellani, Collino, Cozzolino, Curto, Cusimano, Danieli, De Corato, Demasi, Fisichella, Florino, Maggi, Magliocchetti, Magnalb~, Mantica, Marri, Meduri, Monteleone, Mulas, Pace, Palombo, Pasquali, Pedrizzi, Pellicini, Pontone, Ragno, Reccia, Servello, Siliquini, Specchia, Turini e Valentino)

Subject:

NEW SANITARY REGULATIONS AIMING TO ASSURE A RIGHT AND SAFE PRESERVATION OF PRE-PACKAGED FOOD PRODUCTS WITH A SHORT SHELF-LIFE.

Art.1

The Government is delegated to issue (within 18 months from the date this law comes into force) one or more legislative decrees in order to guarantee the protection of consumers' health with special concern in the production, distribution and sale of those food products with a short shelf-life, also considering the existing relation between their production date, their sale and the preservation temperature during their storage.

Art.2

The above-said decrees will have to conform to the leading principles included in the following paragraphs:

Perishability means a biological process causing the loss of original qualities of a food product.

The decrees have to be restricted to fresh and frozen food with a short shelf-life. This one can be extended if the above food products are preserved at low temperatures and they are made and sold in pre-prepared packages.

Shelf-life of a food product means that under special preservation conditions food maintains its original nutritional and structural features and its starting sanitary conditions of preservation. Shelf-life of a food product is the period of time when the food kept at special preservation conditions will maintain nutritional, sensorial features and microbiological quality at a satisfactory level. This happens for a limited time and this period of time is the shelf-life of the product.

"TTI" (Time/Temperature/Integrators) are devices for monitoring thermal history of a food product. "TTI" (Time Temperature Indicators) are devices that give information on a potential time/temperature abuse (according to the preservation condition requirements). They are structured in such a way that they can show evident structural changes in proportion with time-temperature exposition of the food product itself.

Art.3

The above-said legislative decrees to be issued by the Government will eventually modify the following legislative measures still in force: Dlgs 109/92, Dlgs 110/92, Dlgs 155/97, L. 283/62, L. 32/68, L. 690/78, DPR 391/1980, DM 25/9/95 n. 493.

This in order to provide for the obligation to insert, inside and outside the pre-prepared packaging of a food product, one or more TTI with the relevant directions and an additional notice for explaining their final aim.

Art.4

Within three months from the date this law comes into force the Government complies with the obligation provided by paragraph 3 of art.7 included in the EEC Directive 93/43/CEE.

In case the opinion expressed by the foreseen Committee will be favourable, the Italian Ministry of Health in concert with the Ministry of Industry will define (within one year from the date this law comes into force) the distinctive features that TTI must have.

In case the opinion expressed by the foreseen Committee will be favourable, the Italian Ministry of Health in concert with the Ministry of Industry will define (within one year from the date this law comes into force) the performance requirements that TTI must have.

The Ministries will also decide the more suitable setting of TTI in proportion with different kinds of food products on which they have to be applied, even according to the various directions which the food makers apply on their products. Each TTI will have to be calibrated according to the food product shelf-life curve provided by the food processor.

Art.5

The legislative decrees to be issued by the Government will impose specific penalties on the transgressors in conformity with the laws in force concerning safety provisions in the alimentary field.

Art.6

The outline of the above legislative decrees to be issued by the Government is transmitted to the Parliament at least 90 days before the foreseen maturity for the exercise of the delegations. In case the permanent competent Committees don't express their opinion on it within 60 days from the date of the transmission, the decrees can be issued.

Note: This bill was not brought for a vote.

APPENDIX B

CANADIAN OPEN-DATING LEGISLATION
F&D

B.01.007. (1) In this section, "packaging date" means

(a) the date on which a food is placed for the first time in a package in which it will be offered for sale to a consumer; or

19-5-88 (b) the date on which a prepackaged product is weighed by a retailer in a package in which it will be offered for sale for the first time to a consumer.

(1.1) The following information shall be shown on any part of the label:

29-10-92 (a) the identity and principal place of business of the person by or for whom the food was manufactured or produced;

(b) where a prepackaged product having a durable life of 90 days or less is packaged at a place other than the retail premises from which it is to be sold,

(i) the durable life date, and
(ii) instructions for the proper storage of the prepackaged product if it requires storage conditions that differ from normal room temperature; and

19-5-88 (c) where a prepackaged product having a durable life of 90 days or less is packaged on the retail premises from which it is to be sold,

(i) the packaging date, and
(ii) the durable life of the food, except when the durable life appears on a poster next to the food.

(1.2) The packaging date referred to in paragraph (1.1)(c) shall be shown in the form and manner prescribed for the durable life date by subsections (4) and (5) and the terms "best before" and "meilleur avant" on the label shall be replaced by the terms "packaged on" and "empaqueté le".

12-7-79 (2) Paragraph (1.1)(a) does not apply to fresh fruits or fresh vegetables that are prepackaged on retail premises in such a manner that the fruits or vegetables are visible and identifiable in the package.

(3) Paragraph (1.1)(b) and (c) do not apply to

(a) prepackaged products consisting of fresh fruits or fresh vegetables;

(b) prepacked individual portions of food that are served by a restaurant or other commercial enterprise with meals or snacks;

(c) prepackaged individual servings of food that are prepared by a commissary and sold by automatic vending machines or mobile canteens; or

19-5-88 (d) prepackaged donuts.

(4) The durable life date shall be shown in the following manner:

13-5-75 (a) the words "best before" and "meilleur avant" shall be shown grouped together with the durable life date unless a clear explanation of the significance of the durable life date appears elsewhere on the label;

12-3-74 (b) where, for the sake of clarity, it is necessary to show the year in which the durable life date occurs, the year shall be shown first and shall be expressed by at least the last two numbers of the year;

(c) the month shall be shown in words after the year, if the year is shown, and may be abbreviated as prescribed by subsection (5); and

13-5-75 (d) the day of the month shall be shown after the month and shall be expressed in numbers.

(5) The month of the durable life date, when abbreviated, shall be abbreviated as follows and only one such abbreviation shall be used for the English language and the French language:

JA for JANUARY	JL for JULY
FE for FEBRUARY	AU for AUGUST
MR for MARCH	SE for SEPTEMBER
AL for APRIL	OC for OCTOBER
MA for MAY	NO for NOVEMBER
JN for JUNE	DE for DECEMBER

12-3-74

12-3-74 (6) Except as otherwise provided in these Regulations, no person shall use a durable life date marking system on the label of a prepackaged product or in advertising a prepackaged product other than the marking system set out in this section.

12-7-79 (7) Paragraph (1.1)(b) does not apply to prepackaged fresh yeast; if

20-1-77 (a) the date on which it is estimated that the product has lost its effectiveness is shown on the label in the form and manner prescribed for the durable life date by subsections (4) and (5); and

(b) the terms "best before" and "meilleur avant" are replaced by the terms "use by" and "employez avant".

FDR. 1997. Food & Drugs Act & Regulations. Minister of Public Works & Government Services Canada. Ottawa, Canada. Catalog No: H41-1-1997E.

APPENDIX C

NIST Handbook 130

Uniform Open Dating Regulation as adopted by The National Conference on Weights and Measures*

1. Background

Numerous State and local jurisdictions have provided for, or are considering, mandatory open dating of certain packaged commodities. Additionally, many commodities in the marketplace are now voluntarily open dated. Lack of uniformity between jurisdictions could impede the orderly flow of commerce.

In 1985 the National Conference on Weights and Measures, in concert with the Association of Food and Drug Officials, wrote a new Uniform Regulation. It resolved the differences which existed between the versions previously developed by the two organizations independently.

The regulation provides two options for implementation by the States. One requires open dating on all perishable foods. The other permits voluntary open dating of such foods. In the latter (voluntary) case, the open dating must then conform to the uniform regulation. Notes to § 1.1. and 3.1. indicate the alternative wording for the voluntary version of the Regulation.

Uniform Open Dating Regulation[NOTE 1, see page 120]

Section 1. Purpose, Scope, and Application

1.1. **Purpose.**[NOTE 1, see page 120] – The purpose of this regulation is to prescribe mandatory uniform date labeling of prepackaged, perishable foods, and to prescribe optional uniform date labeling that must be used whenever a packager elects to use date labeling on prepackaged foods that are not perishable. Open dating is intended for use and understanding by both distributors and consumers when judging food qualities.

**The National Conference on Weights and Measures is sponsored by the National Institute of Standards and Technology in partial implementation of its statutory responsibility for "cooperation with the States in securing uniformity in weights and measures laws and methods of inspection."*

NOTE 1: *Alternatively, this regulation may be adopted to require uniformity of open dating of perishable foods whenever a packager voluntarily elects to use date labeling. In such instance, Sections 1.1. and 3.1. are reworded in the following manner:*

> ***1.1. Purpose.*** *The purpose of this regulation is to prescribe uniform date labeling that must be used whenever a packager elects to use date labeling on a prepackaged food. Open date labeling is intended for use and understanding by both distributors and consumers when judging food qualities.*
>
> ***3.1. "Sell By" Date.*** *If a retail food establishment elects to sell or offer for sale a prepackaged perishable food identified with a "sell by" date, the "sell by" date used must be as prescribed by this regulation.*

1.2. Scope and Application. – This regulation prescribes the manner of date labeling, the method of determining the appropriate date, required records, responsible persons, and the foods subject to this regulation. This regulation provides for the permissible sale of a regulated food after the expiration of the date on the label. This regulation does not apply to any food that is not prepackaged or is exempted by § 8.

Section 2. Definitions

2.1. "Sell By" Date. – "Sell by" date means a recommended last date of sale that permits a subsequent period before deterioration of qualities described in 2.2., 2.3., and 2.4.

2.2. Perishable Food. – "Perishable food" means any food having a significant risk of spoilage, loss of value, or loss of palatability within 60 days of the date of packaging.

2.3. Semi-perishable Food. – "Semi-perishable food" means any food for which a significant risk of spoilage, loss of value, or loss of palatability occurs only after a minimum of 60 days, but within 6 months, after the date of packaging.

2.4. Long Shelf-life Food. – "Long shelf-life food" means any food for which a significant risk of spoilage, loss of value, or loss of palatability does not occur sooner than 6 months after the date of packaging including foods preserved by freezing, dehydrating, or being placed in a hermetically sealed container.

2.5. **Prepackaged.** – "Prepackaged" means packaged prior to being displayed or offered for retail sale.

2.6. **"Best If Used By" Date.** – "Best if used by" date means a date prior to deterioration of qualities described in 2.3. and 2.4.

2.7. **Person.** – "Person" means an individual, partnership, association, or corporation.

Section 3. Sale of Perishable Food and Date Determination

3.1. **"Sell By" Date.**[NOTE 1, see page 120] – A retail food establishment shall not sell or offer for sale a prepackaged perishable food unless it is identified with a "sell by" date as prescribed by this regulation.

3.2. **Sale after Expiration of "Sell By" Date**

3.2.1. **Advertisement.** – Perishable food shall not be offered for sale after the "sell by" date unless it is wholesome and advertised in a conspicuous manner as being offered for sale after the recommended last date of sale. The placement of a sign, sticker, or tag is acceptable for such advertising if it is easily readable and clearly identifies the perishable food as having passed the recommended last date of sale.

3.2.2. **Responsibility for advertisement.** – The retailer or final seller is responsible for the advertisement, described in § 3.2.1., of a perishable food offered for sale after the recommended last date of sale.

3.3. **Determination of "Sell By" Date**

3.3.1. **Reasonable period for consumption.** – A manufacturer, processor, packer, repacker, retailer, or other person who prepackages perishable food, shall determine a date that allows a reasonable period after sale for consumption of the food without physical spoilage, loss of value, or loss of palatability. A reasonable period for consumption shall consist of at least one third of the approximate total shelf life of the perishable food.

3.3.2. **Responsibility for "Sell By" date.** – A retailer who purchases prepackaged perishable food may upon written agreement with the person prepackaging such food determine, identify, and be responsible for the "sell by" date placed on or attached to each package of such food.

3.4. Manner of Expressing Date

3.4.1. Month and day, or day of week. - A person described in § 3.3.1. or 3.3.2. shall place or attach to each package of perishable food a date by month and day. However, bakery products with a shelf-life of not more than 7 days may be dated with the day of the week representing the last recommended day of sale.

3.4.2. The term "Sell By". - The "sell by" date shall be displayed with the term "sell by" or words of similar import immediately preceding or immediately over the designated date unless a prominent notice is on the label describing the date as a "sell by" date and indicating the location of the date.

3.4.3. Abbreviation of weekday. - If the day of the week is solely designated as provided in § 3.4.1., the name of the day may be abbreviated by the use of either the first two or first three letters of the name of the day.

3.4.4. Expression of month and day. - Except as provided for in § 3.4.1., the date shall be designated by:

(a) the first three letters of the month, preceded or followed by a numeral indicating the calendar day, or

(b) the month represented numerically followed by a numeral designation of the calendar day.

The month and day designation shall be separated by a period, slash, dash, or spacing. When a numeral designation of the first nine days of the month is used, the number shall include a zero as the first digit; for example, 01 or 03. (Amended 1987)

3.4.5. Expression of the year. - The "sell by" date may include the year following the day if such year is expressed as a two or four digit number separated as described in § 3.4.4.

Section 4. Sale of Semi-perishable and Long Shelf-life Food

4.1. **"Best If Used By" Date.** - A manufacturer, processor, packer, repacker, or other person who prepackages semi-perishable or long shelf-life food may place upon or attach to the package an open date providing it is designated by the "best if used by" date.

4.2. **Sale after Expiration of "Best If Used By" Date.** - A retail food establishment may sell or offer for sale food beyond the designated "best if used by" date providing the food is wholesome and the sensory physical quality standards for that food have not significantly diminished.

4.3. **Manner of Expressing Date.** - The "best if used by" date as required by § 4.1. shall be placed upon or attached to each container or package and be limited to the terms "best if used by" or words of similar import followed by or immediately over the date designated by the month and year unless a prominent notice is on the label describing the date as a "best if used by" date and indicating the location of the date. The date shall be designated by the first three letters of the month followed by a numeral indicating the year. The use of the day of the month is permissible provided that the day of the month is placed prior to the month; for example, 30 Jun 81.

Section 5. Placement of the Date

The date, whether "sell by" or "best if used by," shall be printed, stamped, embossed, perforated, or otherwise shown on the package, label on the package, or tag attached to the package in a manner that is easily readable and separate from other information, graphics, or lettering so as to be clearly visible to a prospective purchaser. The date shall not be superimposed on other required information or obscured by other information, graphics, or pricing. Regardless of the type size used, the date shall be easily readable. These requirements do not preclude a supplemental notice elsewhere on a package describing and/or indicating the location of the date.

Section 6. Factors for the Date Determination

A person who, as provided for in this regulation, places either the "sell by" date or "best if used by" date shall determine the date by taking into consideration the food quality, characteristics, formulation, processing impact, packaging or container and other protective wrapping or coating, customary transportation, and storage and display conditions. For purposes of calculating this date, home storage conditions shall be considered to be similar to those in the usual retail store except that the date for refrigerated food may be calculated by using a home storage temperature standard of 40°F (4.4°C).

Section 7. Records

A person responsible for establishing the date for perishable, semi-perishable, and long shelf-life food shall keep a record of the method used for the determination of that date. A record revision is necessary whenever a factor affecting date determination is altered. Such record shall be retained for not less than 6 months after the most recent "sell by" or "best if used by" date and be available during normal business hours for examination upon request by (insert agency name).

Section 8. Exemptions

8.1. This regulation does not apply to perishable fruits or vegetables in a container permitting sensory examination.

8.2. This regulation does not apply to prepackaged perishable foods open dated according to requirements of Federal law or regulation.

Section 9. Preemption of Local, County, and Municipal Ordinance

A municipality or county shall not adopt or impose standards or requirements other than those provided for in this regulation.

Section 10. Effective Date

This regulation shall become effective on and after (insert appropriate date).

APPENDIX D

Extended List of State Regulations on Open Dating
As of September 1998

ALABAMA

Department of Agriculture and Industries Agricultural Chemistry
80-1-22
Rules Governing Foods, Drugs and Cosmetics

80-1-22-.27 Special Requirements for Potentially Hazardous Foods

(1) For the purposes of this Rule, "Potentially hazardous food" means any food that consists in whole or in part of milk or milk products, eggs, or egg products (including inshell eggs), meat, poultry, fish, shellfish, edible crustacea or other ingredients in a form capable of supporting rapid and/or progressive growth of infectious or toxigenic microorganisms.

Statutory Authority: Code of Alabama (1975)
§20-1-2. Adoption of rules and regulations

80-1-22-.28 Special Requirements for Potentially Hazardous Foods Regarding Open Dating

(1) For the purpose of this Rule Open-Date Statement shall mean the terms "Sell by"; "Not To Be Sold After"; "Best If Used By"; "Expiration"; or words of similar import; or a date without additional words shall be considered under the jurisdiction of this Rule.
(2) Potentially hazardous foods as defined under Rule 80-1-22-.27 in package form which are sold, offered for sale, or stored for sale at retail or wholesale and having an open-date statement shall comply with the following requirements.
(3) Prohibitive Acts
    (a) Packages of potentially hazardous foods bearing an open-date statement are not to be sold or offered for sale at retail or wholesale after the date stated on the label.
    (b) Packages of potentially hazardous foods bearing an open date are not to be repacked or relabeled or otherwise altered in a manner that would change the open-date statement originally placed on the package.
    (c) Foods offered for sale in violation of this Rule shall be deemed misbranded under the provision of Code of Alabama, 1975, Section 20-1-25.

Statutory Authority: Code of Alabama (1975)
§20-1-2. Adoption of rules and regulations

Section 20-1-25. When articles deemed misbranded generally—Food.

ALASKA

Voluntary

ARIZONA

ALIS Online
http://www.azleg.state.az.us/cgi-bin/wais...ction=retrieve&WAISdocID=6640713770+0+0+0

http://www.azleg.state.az.us/cgi-bin/wais...tion=retrieve&WAISdocID=6675513826+81+0+0

3-701. Definitions

10. "Expiration date" means the words "sell by" or "buy thru" followed by a date, including the month and day, which is not more than twenty-four days after the eggs were candled and includes the date the eggs were candled.

3-719. Reuse of standard cases and other containers; identification of eggs; expiration date markings; exceptions

E. Cases, half cases, cartons or containers marked grade AA or grade A shall be marked with an expiration **date**. Months shall be abbreviated Jan, Feb, Mar, Apr, May, Jun, Jul, Aug, Sep, Oct, Nov or Dec.
F. The expiration **date** marked on a case, half case or container holding fifteen dozen eggs or more shall be plainly and conspicuously marked in bold-faced type not less than three-eighths inch in height on one outward end of the case or container.
G. The expiration **date** marked on a carton or container holding less than fifteen dozen eggs shall be plainly and conspicuously marked in bold-faced type not less than one-eighth inch in height on one end of the outward top face of each carton and on one outward end or the outward top of each container.
H. Only one description of the size and grade of eggs and one expiration **date** shall appear upon any carton, container, subcontainer or case of eggs.

ARKANSAS

Voluntary

CALIFORNIA

California Annotated Codes (1986)
Div. 15 Milk and Milk Products

§36004. Quality assurance date; listing required on retail packages or containers; exceptions; regulations

(a) At the time of sale to the consumer by a retail store of any product designated in this section there shall appear upon the package or container of such product the date established by the processor as the date upon which, in order to insure quality, such product is normally removed from the shelf or similar location from which the product is offered for sale to the consumer.
(b) This section applies to the following products:
  (1) Market milk.
  (2) Market cream.
  (3) Any milk product which is required by any provision of this code, or by any regulation adopted by the director pursuant thereto, to be made from market milk or any component or derivative of market milk.
(c) This section does not apply to any milk or milk products processed, packaged, and sold by distributors directly to consumers.
(d) This section does not apply to any bulk shipments of milk or milk products between distributors.
(e) The director shall, in compliance with applicable provisions of this code and Chapter 4.5 (commencing with Section 11371) of Part 1 of Division 3 of Title 2 of the Government Code, and after public hearing or hearings, adopt regulations pertaining to (1) responsibility for affixing to packages or other containers the quality assurance date provided for in this section; (2) the manner, style, form, and place of affixation of such date to packages and other containers in a conspicuous place in a form which is readily seen and easily understood by the buyer; and (3) the administration and enforcement of the requirements of this section.

(Added by Stats.1972, c. 703, p.1284, § 1. Amended by Stats. 1974, c. 1155, p. 2452, § 2.)

COLORADO

Voluntary

CONNECTICUT

Connecticut General Statutes Annotated (1985)

§22-197b. Dating of containers of milk, cream, yogurt, cream cheese, cottage cheese, ricotta cheese or sour cream

(a) In addition to the requirements of sections 22-135 and 22-197, each container of milk or cream, yogurt, cream cheese, cottage cheese, ricotta cheese or sour cream sold or offered for retail sale to consumers, on and after January 1, 1982, shall be clearly marked with the last date on which such item may be sold or offered for sale. If such milk or cream was pasteurized at a temperature of two hundred twelve degrees Fahrenheit or less, the last sale date shall not exceed twelve days from the day on which such milk or cream was pasteurized except as provided in subsection (b) of this section.

(b) The milk regulation board shall adopt regulations in accordance with chapter 54[1] establishing a uniform method of displaying such date on such containers and a procedure which the commissioner of agriculture shall follow for approval of a last sale date for milk or cream in excess of twelve days for milk or cream pasteurized at a temperature of two hundred twelve degrees Fahrenheit or less. The regulations shall include but not be limited to procedures for verification of an extended last sale date and review of the appropriateness of such date. The commissioner may authorize an extended last sale date for milk or cream upon request of a milk processor.

(1973, P.A. 73-269, § 1; 1981, P.A. 81-161, § 1, eff. Jan. 1, 1982; 1982, P.A. 82-102, § 1, eff. May 4, 1982; 1984, P.A. 84-19, § 1.)

[1]Section 4-166 et seq.

§ 22-135. Labeling

§ 22-197. Labeling of receptacle containing pasteurized milk or cream

DELAWARE

Voluntary

DISTRICT OF COLUMBIA

District of Columbia Municipal Regulations
June 1997
Title 23: Alcoholic Beverages and Food

2505 Open Dating

2505.1 All pasteurized fluid milk, fresh meat, poultry, fish, bread products, eggs, butter, cheese, cold meat cuts, mildly processed pasteurized products, and potentially hazardous foods sold in food-retail operations which are pre-wrapped and not intended to be eaten on the premises of the food operation shall have easily understood pull dates prominently displayed on their containers.

2505.2 The pull date is the date after which the food may not be sold unless isolated and prominently labelled as being beyond the last date on which the food should be sold without a significant risk of spoilage, loss of value, or loss of palatability if stored by the consumer after that date for the period and in the manner which the food can reasonably be expected to be stored.

2505.3 The display area where the isolated items are kept shall be marked in a manner which will advise the consumer that the goods in the area have been isolated due to the passing of the pull date.

2505.4 The Director shall publish a list, after public hearing, of other foods to be pull-dated.

2505.5 The pull date list established in accordance with §2505.4 shall be reviewed annually and revised as necessary.

2505.6 If any food which has a pull date is rewrapped, the new package shall retain the original pull dates and the package shall be prominently marked with the word "**REWRAPPED.**"

SOURCE: Commissioners' Order 67-1303 effective August 22, 1967, §8-6:106(a)-(c); as amended by Regulation No. 74-1 effective January 19, 1974, §8-6:106(d); 8 DCRR.

FLORIDA

West's Florida Statutes Annotated

502.042. Labeling of shelf life

To ensure consumers full disclosure of the date beyond which milk or milk products may no longer be offered for sale, all dairy processors shall establish, and legibly label as prescribed by rule of the department, the maximum shelf-life period during which milk and milk products may be offered for sale. For purposes of this requirement, to "legibly label" means to label the package or container with conspicuous and easily readable boldfaced print or type in distinct contrast to the background, by color. The department shall periodically conduct shelf-life studies to review the keeping quality of milk and milk products and shall sample periodically the products of the dairy processors to determine if the shelf-life dating used by the processors complies with the minimum standards of quality.

GEORGIA

Rules of Georgia Department of Agriculture
Chapter 40-7-1
General Rules

40-7-1-.26 Labeling.

(2) Open Dating on Food Products in Package Form. All food products in package form and labeled as defined in the Georgia Food Act, Section 26-2-28; which are sold, offered for sale or stored for sale at retail or wholesale and having open dating shall comply with the following requirements. This regulation does not apply to food products which are required to be open dated under other regulations.

(a) Definitions.

1. Open dating. Means the voluntary use of letters (for the month) together with number(s) (for the date of the month) and or a number which designates the last digit of the year.
2. Pull date. The use of a pull date means the last date on which the food in package form shall be sold at retail or wholesale.

(b) Open Date Statement. The terms "Sell by _____"; or "Not To Be Sold After _____"; or "Best If Used By _____"; or Expiration _____" or words of similar import; or a date without additional words shall be considered to fulfill requirements of this regulation.

(c) Manner of Expressing the Pull Date. Open dating shall consist of a combination of three letters (for the month abbreviation) and number(s) for the date of the month; the year may be stated by the last digit of the year. Example: 6-10; or Jun 10; or June 10-8; or 6-10-8.

(d) Location of Stating the Open Date. The open dating information shall be prominently placed on the label with such conspicuousness (as compared with other words, statements, designs, or devices on the labeling) and in such terms as to render it likely to be read and understood by the ordinary individual under customary conditions of purchase and use.

(e) Prohibited Acts. The following acts and the causing thereof are hereby prohibited:

1. Food products in package form are not to be sold or offered for sale at retail or wholesale after the expiration date stated on the label.
2. Food products in package form shall be removed from sale after the pull date stated on the label. These food products may be voluntarily destroyed

or returned to the manufacturer, packer or distributor provided the food shows no evidence of spoilage and is not suspected of being a health hazard.

3. The alteration, mutilation, destruction, obliteration, or removal of the whole or any part of the labeling of, or the doing of any other act with respect to a food if such act is done while such article is held and results in such product being adulterated or misbranded.

(3) Expiration Date of Infant Formula Products.

(a) Infant Formula. Expiration Dating.

1. Each and every container of liquid or powdered infant formula made from two or more ingredients and represented as or intended as a replacement or supplement for milk, shall conspicuously show in common and express terms the calendar month and year after which the product is not to be sold or used for human consumption.

2. The expiration date, or the date after which the product is not to be sold or used for human consumption, shall be determined by the manufacturer based on empirical data, or other verifiable scientific means.

3. If the Commissioner or his authorized agent has prohibited the sale of a product still within date, after notice, the manufacturer shall for each and every brand, variety, or formulation of infant formula intended to be sold or offered for sale in Georgia, submit scientific data establishing the expiration date to comply with Section (a) 2 of this regulation. Such data shall include, but is not limited to, physical, nutritional, and chemical properties. In the absence of empirical data on any specific formulation, the manufacturer shall provide such scientific data to reasonably substantiate the expiration date. If the data submitted does not in the opinion of the Commissioner justify the expiration date, the Commissioner shall prohibit the sale of the product until a new expiration date consistent with data is applied to the food product.

4. Each and every shipping carton container, and consumer package shall in like manner show the calendar month and year after which the product is not to be sold or used. This section will be complied with if the information is not contained on the shipping carton, container, and consumer package, but is easily legible by virtue of the transparency of the outer wrapper or container.

5. Any manufacturer, distributor, dealer, or other person who offers for sale or sells infant formula not showing an expiration date or who offers for sale or sells infant formula on a date after the expiration date shown, shall be deemed to be offering for sale a product unfit for food within the meaning of part 26-2-28 of the Georgia Food Act.

6. Knowingly filing with the Commissioner incorrect or unverifiable data or placing an expiration date upon a shipping carton, container or any

consumer package, which date is inconsistent with the data filed with the Commissioner shall be deemed to be misbranding under part 26-2-28 of the Georgia Food Act provided, however, that it shall not be deemed misbranded if the expiration date shown is an earlier date than the filed data would warrant.

(b) Special Formulation. The provisions of this regulation shall not apply to any special formulation manufactured on request of any licensed physician for the express purpose of meeting dietary needs of a specific individual.

(4) Sandwiches.

(a) For the purpose of this section, prepackaged sandwiches shall be classified as Type A, Type B, or Type C.

1. Type A Sandwiches are those prepackaged sandwiches which are handled and sold as non-refrigerated sandwiches. Type A sandwiches shall not contain foods defined as potentially hazardous in hold or as an ingredient.
2. Type B Sandwiches are those prepackaged sandwiches which are handled and sold as refrigerated sandwiches.
3. Type C Sandwiches are those prepackaged sandwiches which are immediately hard frozen after manufacture and which are kept at zero degrees Fahrenheit until the time of sale to institutions or individual retail firms. A 10 degree rise in temperature will be allowed during delivery of frozen sandwiches from the truck into an individual firm.

(b) Type A Sandwiches which are stored, transported and offered for sale in a non-refrigerated state shall be labeled with a conspicuous expiration date not later than two (2) days from the date of manufacture.

(c) Type B Sandwiches which are labeled "Keep Refrigerated" and which are stored, transported and offered for sale under refrigeration shall be held after preparation continuously at or below 40 F, and shall be labeled with a conspicuous expiration date not later than fourteen (14) days from the date of manufacture.

(d) Type C Sandwiches shall be dated when sold by a retail store account for resale to individual customers, the expiration date, not to exceed fourteen (14) days, shall be stamped on each individual sandwich label when the sandwich is removed from the freezer, placed under refrigeration at or below 40 F, and made ready for sale to individual customers. The shelf life date of the sandwich shall be preceded by a statement: "Not to be sold after (date)"; or words of similar intent. If sandwiches are intended to be maintained in a non-refrigerated state after removal from the freezer, a conspicuous expiration date not to exceed two (2) days from the date of such removal shall be stamped on each individual sandwich. The date at the end of the shelf life of the sandwich shall be preceded by a statement: "Not to be sold after (date)"; "Sell By (date)"; or words of similar intent.

(e) Type B & C Sandwiches intended to be sold as refrigerated sandwiches shall be labeled "Keep Refrigerated".
(f) The date shall be conspicuously located on the front of the wrapper.
(g) The expiration date for sandwiches shall state the last day of the sale in terms of the month or its abbreviation and numerical day of the month (e.g. 6-6). The expiration day shall be preceded by an explanatory term, such as "Expires", "Sell-By", or similar wording. Other codes or dating methods are prohibited.
(h) When improved packing methods including but not limited to "Nitrogen Flushed" and "Vacuum Packed" are developed and become economically feasible, an extended expiration date may be granted by the Commissioner upon application by the manufacturer with proof of the safety of such an extended shelf life. The time of such extension will be decided by the Commissioner based upon such proof from the manufacturer.

Authority Ga. L. 1956, p. 195 as amended; O.C.G.A. Sec. 26-2-1 et. Seq

Georgia Food Act
Section 26-2-28. When food deemed misbranded.

HAWAII

Hawaii Administrative Rule
Title 11. Department of Health
Chapter 15. Milk

§11-15-39 Labeling.

(c) Every container of processed milk and milk product held in retail and wholesale stores, restaurants, schools, or similar establishments for sale shall be conspicuously and legibly marked by the milk plant with the designation of the month and day of the month after which the milk shall not be sold for human consumption. If the director determines that the quality of a milk or milk product deteriorates to the extent that the product becomes unmarketable prior to its designated shelf-life, the director shall notify the milk plant to conduct a full and complete review of the shelf-life of the product. If the review confirms that the designated shelf-life of the product is improper, the milk plant shall immediately shorten the designated shelf-life of the product in order to maintain the quality of the product throughout its intended shelf-life.

IDAHO

Voluntary

ILLINOIS

Voluntary

INDIANA

Voluntary

IOWA

Inspections and Appeals[481]
Chapter 31. Food Establishment Inspections

481—31.11(137A) Reduced oxygen packaging of food. All places which engage in reduced oxygen packaging: cook-chill, vacuum packaging, sous vide, modified atmosphere packaging, or controlled atmosphere packaging, shall comply with standards provided by the Association of Food and Drug Officials Retail Guidelines: Refrigerated Foods in Reduced Oxygen Packages of 1990. Existing reduced oxygen packaging operations will have until January 1, 1996, to comply with this rule. This rule is intended to implement Iowa Code sections 137 A.5, 137B.3 and 137B.6. These rules are intended to implement Iowa Code chapter 137A...

Association of Food and Drug Officials
Retail Guidelines
Refrigerated Foods in Reduced Oxygen Packages

LABELING—"USE BY" DATES

Each package of refrigerated retail processed food in a reduced oxygen atmosphere must bear a "use by" date. This date cannot exceed 14 days from retail processing. Also, the date assigned by the retailer cannot go beyond the manufacturer's recommended "Pull Date" for the food. The "use by" date must be listed on the principal display panel in bold type on a contrasting background. Foods that remain frozen before, during, and after processing are exempt from this requirement.

KANSAS

Voluntary

KENTUCKY

OPEN DATING REQUIREMENTS
For Milk Products
Issued by the Cabinet for Human Resources
Department for Health Services
Division of Consumer Health Protection
Milk Control Branch
(902 KAR 50:080)

Relates to KRS 217.010 to 217.990
Pursuant to KRS 13.082, 194.050, 211.090(1)(c)
Necessity and Function: The Cabinet for Human Resources is directed by KRS Chapter 217C to regulate the production, transportation, processing, handling, sampling, examination, grading, labeling, standards of identity, sale and such other matters relating to milk and milk products as may be necessary to protect the public health. This regulation establishes uniform standards for the open dating of Grade A pasteurized milk and milk products in Kentucky and requires that a consumer information or purchase guide date be legibly applied to retail packages offered for sale to the final consumer.

Supersedes: MMP-6
Section 1. Open Date Required.

No person shall sell or offer for sale any Grade A pasteurized milk or milk product in this state in a consumer package that does not bear the open date as required by this regulation.

Section 2. Open Date Labeling.

(1). The open date shall appear in such form as to be conspicuous, legible and commonly understandable. It shall be boldface print in contrast to the background, by typography, color, embossing, debossing or molding of other matter on the package, and shall be placed on that part of the container most likely to be displayed, presented, or shown, or examined under customary conditions of display for retail sale, and shall not interfere with legibility of other mandatory labeling requirements of the product.
(2). The open date shall be expressed by the first three (3) letters of the month followed by or preceded by the numeral or numerals constituting the appropriate calendar date, or expressed numerically by the number of the month preceding the number of the day. (For example: June 1 may be expressed "Jun 1", "1 Jun", "06 01" or "06-01.")

(3). The open dating of Grade A pasteurized milk and milk products bottled in glass containers for home delivery are exempt from this regulation.

Section 3. Certification of Open Date Required.

(1). Each distributor or processor manufacturing, processing or packaging Grade A pasteurized milk and milk products for sale within this state shall comply with the provisions of this regulation and file and certify with the cabinet the open date for each product. The processor or distributor requesting an open date certification for a product shall provide the cabinet with the research data used to support the product open date certification request.
(2). Samples of products for open date evaluation may be obtained at the processing plant, delivery truck, distributors or from retail outlets. The temperature and open date shall be officially reported at the time of sample collection.

Section 4. Enforcement.

In the event a product is not sold within the period specified in the open date, the cabinet shall take appropriate action necessary to remedy the condition consistent with applicable public health laws and regulations.

EFFECTIVE DATE: January 8, 1975

LOUISIANA

Voluntary

MAINE

Voluntary

MARYLAND

Annotated Code of Maryland (1996)
1997 Supplement
Subtitle 4. Milk Products.
Part III. Standards and Handling Requirements for Milk Products

§ 21-426. Grade A milk product dating.
(a) *"Sell-by period" defined.* — In this section, "sell-by period" means the length of time a Grade A milk product may be kept for sale.

(b) *Container markings.* — (1) After a public hearing, the Secretary shall adopt rules and regulations that establish a method for conspicuously marking, on the Grade A milk product container, the last date on which the product may be sold.

(2) Each Grade A milk product container shall be marked as required by rules and regulations adopted under this subtitle.

(c) *Sell-by period established.* — A Grade A milk product that is cooled to, packaged, and stored at 45 degrees Fahrenheit or less before it is purchased by or delivered to the ultimate consumer has a sell-by period of 14 days.

(An. Code 1957, art.43, § 574; 1982, ch.240, § 2; 1987, ch.306, § 2; 1989, ch.461; 1996, ch.691.)

Effect of amendments. - The 1996 amendment, effective June 1, 1996, deleted former (b), (d)(2) and (e); redesignated former (c) as present (b) and former (d)(1) as present (c); and substituted "14 days" for "7 days" in present (c).

MASSACHUSETTS

The Commonwealth of Massachusetts
105CMR 500.000-529.000

520.119: Food, Open Date Labeling

(A) Purpose. The purpose of 105 CMR 520.101 through 520.205, is to cause certain food products offered for sale in the Commonwealth of Massachusetts to be identified relative to physical sensory qualities for the information of consumers, distributors, and retailers.

(B) Scope. 105 CMR 520.101 through 520.205 prescribe the method of posting an identification date, date determination, required records, responsible persons and foods subject to 105 CMR 520.119. In addition, 105 CMR 520.101 through 520.205 provide for exemption of certain foods and for sale of foods, under stipulated conditions, after the expiration of an identifying open date.

(C) Definitions. For purposes of 105 CMR 520.101 through 520.205, the following definitions shall be applicable:

Best If Used by Date: A date no later than the expiration of the estimated shelf life of a food product.

Frozen Food: A food product which has been packaged and preserved by freezing.

Long Shelf-Life Food: A food product having an estimated shelf life of 90 days or more, including foods preserved by dehydration or packaged in a hermetically sealed container, but excluding frozen foods.

Perishable Food: A food product having an estimated shelf life of 60 days or less.
Person: An individual, partnership, association, or corporation.
Prepackaged: Packaged prior to being displayed or offered for sale at retail.
Sell by Date: A recommended last date of retail sale of a food product which provides for a reasonable subsequent period of home shelf life.
Semi-Perishable Food: A food product having an estimated shelf life greater than 60 days but less than 90 days.
Shelf Life: A period of time after the date of packaging during which a food product has no significant risk of spoilage, loss of value, or loss of palatability, given compliance with recommended conditions of storage and handling as disclosed on the label of such product.

(D) Open Dating of Perishable and Semi-Perishable Food Products. No person shall sell, offer for sale, or have in his possession with intent to sell, prepackaged perishable or semi-perishable food products unless they are identified with a "sell-by-date" or a "best if used by date" determined by manufacturer, processor, packer, repacker, retailer, or other person who had packaged such food products and displayed in the form specified in 105 CMR 520.119(G).

(E) Open Dating of Frozen and Long Shelf Life Food Products. A manufacturer, processor, repacker, retailer, or other person who prepackages frozen or long shelf life food products may mark the individual retail packages of such products with a "sell by date" or a "best if used by date," but shall be subject to the requirements as to form outlined in 105 CMR 520.119(G).

(F) Sale of Past Date Food Products. No person shall offer for sale in the Commonwealth any food product after the expiration of a "sell by date" or a "best if used by date" unless:

(1) It is wholesome and its sensory physical qualities have not significantly diminished; and,
(2) It is segregated from food products which are not "past date"; and,
(3) It is clearly and conspicuously marked either on the package or through the use of shelf markers or placecards, as being offered for sale after the recommended last date of sale or best use.

(G) (1) Placement of the Date. A date shall be displayed with the term "sell by" or "best if used by" in reasonable proximity to the designated date.

(2) Such a date shall consist of the common abbreviation for the calendar month and numerals for the day and year, e.g., Feb. 10, 1980; or numerals for the month, day and year, e.g., 2/10/80, except that:

(a) Perishable food products need not have the year identification included in the date, and frozen and long shelf life foods need not have the day identification included in the date.

(b) Fresh bakery products may be dated with only the day designation, e.g., Monday, or an abbreviation thereof, e.g., Mon.

(3) A date shall be accompanied by disclosure of recommended product storage conditions, if such conditions significantly affect the validity of such a date.

(4) A date and any recommended storage conditions shall be printed, stamped, embossed, perforated, or otherwise shown on the retail package, a label on such package, or a tag attached to such package in a manner that is easily readable and separate from other information, graphics, or lettering so as to be clearly visible to a prospective purchaser.

(5) If a date and recommended storage conditions do not appear on the principal display panel, the information panel, or on another conspicuous portion of the individual retail package, a statement must appear on the principal display or information panel indicating where such information can be found elsewhere on the package.

(6) An individual prepackaged food product which is not labelled in accordance with the provisions of 105 CMR 520.119 shall be deemed "mis-branded" pursuant to M.G.L. c. 94, § 187.

(H) Product Rotation, Storage and Handling Information. Any person who prepackages a food product for sale in the Commonwealth shall disclose to the retailer of such product;

(1) Whether or not such product is open dated

(2) Any required or recommended storage or handling conditions.

(3) Information to facilitate the sequential rotation of product inventory. Information shall be conveyed in a readily understandable form.

(I) (1) Factors for Shelf-Life Determination. A person who is responsible for placing a "sell by date" or a "best if used by date" on a food product as required or permitted by these regulations, shall estimate the shelf life of such a product, taking into consideration the quality, characteristics, formulation, processing impact, packaging or container and other protective wrapping or coating, typical transportation and storage and display conditions of such a food product.

(2) Considerations should also include those of the retail store and consumer; and for purposes of estimating shelf life, home storage conditions shall be considered similar to the usual retail store, except that refrigerated food may be calculated using a home temperature storage standard of 40 ¼ °F or 4.4 ¼ °C.

(3) Such factors shall be measured or otherwise determined utilizing testing and sampling procedures customarily utilized by the food industry for such purposes.

(J) (1) Required Records. A person responsible for estimating the shelf life of a food product shall keep a record of the method used for the determination of such shelf life and the corresponding "sell by date" or "best if used by date." A record revision is necessary whenever a factor affecting such date determination is altered. Such record shall be retained for not less than six months after the most recent "last date of sale" or "best if used by date" and be available upon written request by the Department.

(2) If, after conducting an investigation, the Department determines that the date selected is not supported by such records, it may direct the responsible person to change the date in accordance with such findings. Any person aggrieved by such an order shall be afforded the opportunity for hearing; but the order shall not be overturned unless the appellant establishes by clear and convincing evidence that the date originally selected is, in fact, justifiable. The order shall be considered final unless reversed upon such review; and pending review, the affected product may not be offered for sale, unless the date is modified in accordance with the order of the Department.

(K) (1) Exemptions. 105 CMR 520.101 through 520.205 do not apply to:

(a) Fresh meat, fresh poultry, fresh fish, fresh fruits, and fresh vegetables offered for sale unpackaged or in a container permitting sensory examination.

(b) Salt and crystallized refined sugar.

(c) Food products shipped in bulk form for use solely in the manufacture of other foods and not for distribution to the consumer in such bulk form or container.

(d) Individually packaged food products which are prepackaged as components of a larger food item, if the larger food item is identified with a date no later than the corresponding date for any such components.

(e) Food products prepackaged for retail sale with a net weight of less than 1½ ounces.

(f) Food products manufactured for sale outside the Commonwealth, processed for sale outside the Commonwealth, or stored for sale outside the Commonwealth.

(2) Any person may apply to the Department for an exemption from the provisions of 105 CMR 520.000, which exemption shall be granted if the product for which the exemption is sought:

(a) Is open dated in accordance with the regulations of another agency; and, compliance with the regulations of another agency; and, compliance with the regulations of the other agency will result in the disclosure of substantially the same information as is required by 105 CMR 520.000; or,

(b) Has been voluntarily open dated prior to the promulgation of 105 CMR 520.000 in a manner which will result in the disclosure of substantially the same information as is required by 105 CMR 520.000.

(L) Effective Dates. Notwithstanding any other effective dates set forth in 520.119(L) shall take effect in accordance with the following schedule:
   (1) For perishable food, 105 CMR 520.119(C), the effective date shall be the date on which 105 CMR 520.000 are published March 12, 1981.
   (2) For semi-perishable foods, 105 CMR 520.119(C), the effective date shall be May 1, 1982.
   (3) For frozen foods, 105 CMR 520.199(C), the effective date shall be December 1, 1981.
   (4) For long shelf life foods, 105 CMR 520.119(C), the effective date shall be May 1, 1982.

MICHIGAN

Department of Agriculture
Dairy Division
Regulation No. 408. Fluid Milk and Milk Products
http://www.state.mi.us/execoff/admincode/data/AC00285/s00408/.txt

(By authority conferred on the department of agriculture by section 1 of Act No.233 of the Public Acts of 1965, as amended, being S288.21 of the Michigan Compiled Laws)

R 285.408.1 Definitions.

Rule 1. (1) "Container" includes the closure.
(2) "Department" means the department of agriculture.
(3) "Date" means the recommended last day of sale.
(4) "Milk and milk products" means and includes cream, light cream, coffee cream, table cream, whipping cream, heavy cream, heavy whipping cream, whipped cream, whipped light cream, whipped coffee cream, whipped table cream, sour cream, cultured sour cream, half-and-half, cultured half-and-half, reconstituted or recombined milk and milk products, concentrated milk and milk products, skim milk, skimmed milk, nonfat milk, lowfat milk, fortified milk and milk products, vitamin D milk and milk products, homogenized milk, flavored milk and milk products, buttermilk, cultured milk, cultured whole milk buttermilk, yogurt, lowfat yogurt, acidified milk and milk products, grade A cottage cheese, grade A lowfat cottage cheese, grade A dry curd cottage cheese, grade A egg nog, and other milk products which may be defined in the future under Act No, 233 of the Public Acts of 1965, as amended, being S288.21 et seq. of the Michigan Compiled Laws.

History: 1954 ACS 89, Eff. Oct. 21, 1976; 1979 AC.

R 285.408.2 Last day of sale.

Rule 2. (1) Each processor and manufacturer of milk and milk products sold in this state shall place on each container of milk and milk products a recommended last day of sale by month and day.

(2) The date shall be expressed by the first 3 letters of the month followed by the numeral designating the appropriate calendar day, or by designating the calendar month numerically followed by a numeral designating the calendar day. For example, September 9 may be expressed as "Sep. 09" or "9.09." When the numerical expression is used, a day numbering less than 10 shall be shown with a zero as the first digit; for example 01 or 03.

(3) The date shall appear on that part of the container that is most likely to be displayed, presented, or shown under customary display conditions of sale. However, a cup container may have the date placed on the bottom.

(4) The date on the container shall be legible and shall not interfere with the legibility of other information required to be on the product.

History: 1954 ACS 89, Eff. Oct. 21, 1976; 1979 AC.

R 285.408.3 Time interval of date.

Rule 3. Processors and manufacturers of milk and milk products shall register the following information with the department on forms provided by the department:

(a) The assigned date of each milk and milk product processed and the length of time between production and the date. This length of time shall be substantiated by plant records of a testing program conducted by the processor or manufacturer.

(b) The method of application, and location of the date, for each size and style of container.

(c) Changes in the time interval of the date prior to the effective day of the change.

History: 1954 ACS 89, Eff. Oct. 21, 1976; 1979 AC.

R. 285.408.4 Flavor.

Rule 4. (1) Milk and milk products shall maintain nutritional levels and shall not have a flavor change before the date.

(2) The department shall periodically sample and analyze milk and milk products to determine if the flavor has changed by the date. Milk and milk products obtained for analysis by the department prior to the date shall be stored at a temperature of 44 degrees Fahrenheit, plus or minus 1 degree, until analyzed.

(3) The processor or manufacturer of milk or milk products which do not maintain their flavor until the date shall, upon receipt of written or verbal notice from the department, make the changes necessary to improve product quality or alter the date so as to comply with these rules. The processor manufacturer is not responsible for milk and milk products when the nutritive value loss or flavor deterioration of those products can be determined to be caused by mishandling, improper storage, or lack or refrigeration at points beyond his control.

History: 1954 ACS 89, Eff. Oct. 21, 1976; 1979 AC.

R 285.408.5 Sale after date.

Rule 5. (1) Milk and milk products shall not be offered for sale after the date, unless they are advertised to the final consumer in a prominent manner as being beyond the recommended last day of sale.

(2) The final seller is fully responsible for the proper advertisement of milk and milk products sold beyond the date.

History: 1954 ACS 89, Eff. Oct. 21, 1976; 1979 AC.

MINNESOTA

Minnesota Statutes Annotated (1996)
Quality Assurance Dating

31.781 Declaration of Policy.

The legislature recognizes the entire food industry in the nation as leaders in the world in providing wholesome, nutritious, fresh and clean food to its citizens and to others. The Minnesota department of agriculture is hereby authorized and directed to promulgate rules which provide for a quality assurance date on perishable foods, to assure this industry's continuation and the degree of improvement reasonable and feasible, so as to provide people with wholesome, nutritious, fresh and clean food.

History: 1973 c 686 s 1; 1985 c 28 s 70

31.782 Definitions.

Subdivision 1. As used in sections 31.781 to 31.789, the following terms shall have the meaning ascribed to them.

Subd. 2. [Repealed, 1996 c 310 s1]

Subd. 3. "Perishable food" means any food intended for human consumption (other than meat and poultry, frozen food, or fresh fruit or vegetables), which has a quality assurance date.

Subd. 4. "Quality assurance date" means any date after which the manufacturer or processor reasonably determines that the product may, by spoilage, wiltage, drying or any other foreseeable and normal natural phenomenon, lose its palatability or its desired or nutritive properties. The date shall include the day, month, and if appropriate, the year.

History: 1973 c 686 s 2

31.783 Rules, Scope.

Subdivision 1. The commissioner shall administer and enforce the provisions of sections 31.781 to 31.789 by rules adopted prior to October 1, 1973 pursuant to the administrative procedure act.

Subd. 2. Perishable foods which bear a quality assurance date of 90 days or less from the date of packaging shall be dated in accordance with the rules adopted pursuant to sections 31.781 to 31.789.

Subd. 3. Perishable foods which bear a quality assurance date of more than 90 days from the date of packaging may require dating in accordance with rules adopted pursuant to sections 31.781 to 31.789.

Subd. 4. Whenever the commissioner has reason to believe that any rule adopted pursuant to sections 31.781 to 31.789 is inappropriate or unsuitable to any particular perishable food product or products, the commissioner may, in accordance with the administrative procedure act, waive the application of such rules as to such product or products.

History: 1973 c 686 s 3; 1985 c 248 s 70

31.784 Expiration of Quality Assurance Date.

Nothing contained in sections 31.781 to 31.789 or any rule adopted pursuant hereto shall require the removal from sale of the perishable food product after the expiration of the quality assurance date on the product nor imply that after the expiration of the quality assurance date on the product, the product is not wholesome or safe for human consumption.

History: 1973 c 686 s 41 1985 c 248 s 70

31.785 Regulations of Other States and the Federal Government.

If any other state, or the federal government, adopts an open dating statute or regulation which provides for information and enforcement equal to or greater than that of sections 31.781 to 31.789, the commissioner may, by rule, exempt any product from the provisions of sections 31.781 to 31.789 if it is in compliance with such other statute or regulation.

History: 1973 c 686 s 5; 1985 c 248 s 70

MISSISSIPPI

Mississippi Egg Marketing Law and Regulation
Mississippi Department of Agriculture and Commerce

Regulation No. 1 — Date beyond which eggs may not be sold.

All eggs offered for sale in consumer packages (cases, boxes, baskets, or containers) shall be legibly dated (month and day or consecutive day of the year) the day the eggs were packed and may bear an expiration date of no more than (30) days from date of pack, excluding date of pack. Provided, however, notwithstanding a labeled expiration date, all eggs subject to the provisions of the Mississippi Egg Marketing Law shall not be offered for sale or sold when thirty (30) days shall have lapsed since the eggs were packed, excluding date of pack.

Shell eggs labeled AA or Fresh Fancy shall bear in distinctly legible form an expiration date of no more than ten (10) days from date of pack, excluding date of pack. The expiration date shall be stated as the "month" and "day" (i.e., April 3 or 4-3) preceded by the letters EXP or SELL BY. (Adopted March, 1988)

MISSOURI

Voluntary

MONTANA

Administrative Rules of Montana
Fluid Milk and Grade A Milk Products
Sub-Chapter 2
Milk Freshness Dating

32.8.201 Scope of Rules

(1) These rules apply to whole milk, low fat milk, nonfat milk, chocolate milk, whipping cream, half and half and/or any other liquid milk product designed to be consumed in the form in which it is packaged, except

buttermilk. For purposes of this sub-chapter "milk" means any of the above products.

History: Sec. 81-2-102 MCA; IMP, Sec. 81-2-102 MCA; NEW, 1980 MAE p. 1603, Eff. 7/1/80.)

32.8.202 Time from Processing that Fluid Milk may be Sold for Human Consumption

(1) No grade A pasteurized milk may be sold, offered for sale, or otherwise disposed of for human consumption at retail or wholesale more than 12 days after pasteurization is completed.
(2) No grade A raw milk may be sold, offered for sale, or otherwise disposed of for human consumption at retail or wholesale more than 12 days after the bottling is completed.
(3) For purposes of this rule, the 12 day period ends on the first midnight following 12 consecutive 24 hour days. In no instance may the period be less than 288 hours.
(4) No grade A pasteurized milk or grade A raw milk may be put in any container marked with a pull date which is more than 12 days after pasteurization or bottling of the milk, whichever is applicable, without notification to the department of both the amount of and state destination of the milk. The department will provide necessary forms for detailing the amount and destination of the milk.
(5) No grade A pasteurized milk or grade A raw milk put in any container marked with a pull date which is more than 12 days after pasteurization or bottling of the milk, whichever is applicable, may be offered for sale or otherwise disposed of for human consumption at retail or wholesale in Montana.
(6) Unless otherwise agreed upon, the person who offers the milk for sale to the public is responsible for removing the milk at the expiration of the 12 days.

History: Sec. 81-2-102 MCA; IMP, Sec. 81-2-102 MCA; NEW, 1980 MAR p. 1603, Eff. 7/1/80; AMD, 1986 MAR p. 50, Eff. 1/17/86; AMD, 1987 MAR p. 698, Eff. 5/29/87.

32.8.203 Labeling of Milk Containers to Show Last Day of Legal Sale

(1) Each container into which grade A pasteurized or grade A raw milk is placed for sale for human consumption must be marked with a pull date. The date will state in arabic numerals or standard abbreviations for months, the month and day which is the last day the milk may be sold as set forth in ARM 32.8.202.

(2) Language in substance the same as "sell by" or "not to be sold after" must be placed by the date in a manner which clearly shows that the milk must be sold by the date on the container.

(3) No person, other than the packager of the milk, may mark the package with a pull date without permission of the department of livestock.

History: Sec. 81-2-102 MCA; IMP, Sec. 81-2-102 MCA; NEW, 1980 MAR p. 1603, Eff. 7/1/80; AMD, 1987 MAR p. 698, Eff. 5/29/87)

32.8.204 Exemption from Labeling Requirement

(1) Licensed grade A raw milk dairies are exempt from the labeling requirements imposed by ARM 32.8.203 when all milk packaged for human consumption is sold directly to the consumer either at the licensed retail raw dairy or through a delivery route directly operated by the licensed retail raw dairy.

History: Sec. 81-2-102 MCA; IMP, Sec. 81-2-102 MCA; NEW, 1980 MAR p. 1603, Eff. 7/1/80.)

32.8.205 Manner, Positioning, and Size of Labeling

NEBRASKA

www.agr.state.ne.us

(66) **Reduced Oxygen Packaging**

(a) "Reduced oxygen packaging" means the reduction of the amount of oxygen in a PACKAGE by mechanically evacuating the oxygen; displacing the oxygen with another gas or combination of gases; or otherwise controlling the oxygen content in a PACKAGE to a level below that normally found in the surrounding atmosphere, which is 21% oxygen.

(b) **"Reduced oxygen packaging"** includes methods that may be referred to as altered atmosphere, modified atmosphere, controlled atmosphere, low oxygen, and vacuum PACKAGING including sous vide.

81-2,272.27. Food establishment; reduced oxygen packaging method; when.

(2) A food with a high level of competing organisms such as raw meat, raw poultry, or semi-soft cheese containing live active starter culture organisms may be packaged using a reduced oxygen method. Such products shall be labeled with a "sell by" or "use by" date not to exceed fourteen days and shall be discarded if not sold by that date.

(4) Except as provided in subdivision (2) of this section, products packaged using a reduced oxygen method shall be discarded if not sold within thirty days from processing if the food is processed at the food establishment. Food processed by a food processing plant that has been repackaged by the food establishment shall be discarded if not sold within fourteen days from packaging by the food establishment or the original manufacturers "sell by" or "use by" date, whichever occurs first.

(5) Except as provided in subdivision (2) of this section, products packaged using a reduced oxygen method shall be labeled with a "sell by" or "use by" date which is not more than thirty days from the date it was processed by the food establishment. Food processed by a food processing plant that has been repackaged by the food establishment shall be labeled with a "sell by" or a "use by" date which is not more than fourteen days from the date it was repackaged by the food establishment or the original manufacturer's "sell by" or "use by" date, whichever occurs first.

Nebraska Graded Egg Act and Rules and Regulations

2-3509 Shell eggs; sold without designation of date packed; unlawful.

It shall be unlawful to sell shell eggs in any carton or container which fails to show the date of the month on which the egg were packed.

NEVADA

Voluntary

NEW HAMPSHIRE

New Hampshire Department of Agriculture, Markets & Food
Weights and Measures Rules

Chapter 1400
PART Agr 1413
Dating Prewrapped Sandwiches

Agr 1413.03 Definitions. The following words shall have the meanings indicated when used in this chapter:

(c) "Expiration date" means the last day of sale, printed or stamped on a prewrapped sandwich, determined in accordance with these rules.

(f) "Prewrapped sandwich" means any sandwich containing meat, poultry, seafood, vegetables, cheese or mayonnaise, enclosed in a wrapper, stored in a refrigerator for future consumption and offered for sale or sold to the public, which is not intended or expected to be consumed within 36 hours after wrapping. This does not include sandwiches which are sterilized and vacuum packed.

Agr 1413.04 Expiration Date

(a) The expiration date for a refrigerated prewrapped sandwich shall be a maximum of 7 days from the date the sandwich is made, but can be less than 7 days, as determined by the vendor in accordance with 1413.04 (d) & (e).
(b) The expiration date for a frozen prewrapped sandwich shall be a maximum of 14 days from the date the sandwich is made, but can be less than 14 days, as determined by the vendor in accordance with 1413.04 (d) & (e).
(c) The expiration date shall be printed or stamped on the sandwich wrapper, at the time it is wrapped, by stating "expiration date" or "sell by" followed by the month and day.
(d) The expiration date shall be determined by the vendor who makes the prewrapped sandwiches. The vendor shall determine a date which allows a reasonable period after sale for consumption of the food without physical spoilage. This determination shall take into consideration the food quality and characteristics, packaging and other protective wrapping, coating or container, transportation and storage conditions, and retail storage and display conditions. The temperature and humidity of storage and display areas for perishable food, at all stages prior to retail sale, shall also be factors in determining the expiration date.
(e) The determination of a reasonable period for consumption shall be based upon refrigeration temperatures of less than 40 degrees Fahrenheit, or 4.4 degrees Celsius or, if frozen, less than 0 degrees Fahrenheit, or -17.8 degrees Celsius.
(f) A retailer who purchases prewrapped sandwiches may, upon written agreement with the vendor, determine, identify and be responsible for the date placed on each prewrapped sandwich.
(g) The person responsible for establishing the date for perishable food shall keep a record of the method of determination of the date. Such record shall be retained for at least 6 months.

Agr 1413.05 Complaint

A complaint may be filed by any person, in accordance with the following provisions:
(a) A complaint shall be sent by first class mail to the department of agriculture, containing the following:
(1) The complainant's name and address;
(2) The alleged violator's name and address;

(3) The specific fact and circumstances which constitute a violation shall be as follows:
(a) No stated expiration date on a prewrapped sandwich; or
(b) The prewrapped sandwich is for sale past the expiration date.
(4) The signature of the complainant.

Agr 1413.06 Investigation.

Upon receipt of a complaint, the commissioner shall inspect the alleged violator's premises, in accordance with Agr 1413.07, to determine whether there is or was a violation of RSA 438:26-b, or a rule or order of the commissioner.

Agr 1413.07 Right of Inspection.

(a) The commissioner shall enter the premises of any vendor or retailer, stop any person on the premises and seize any packages sold in violation of the law, without a formal warrant, pursuant to RSA 438:15.
(b) The commissioner shall inspect the premises of a vendor or retailer at any time during business hours to determine whether there is a violation of RSA 438:26-b or a rule or order issued thereunder.

Agr 1413.08 Orders.

(a) The commissioner shall issue an order when there is no stated expiration date on a prewrapped sandwich or the prewrapped sandwich is for sale past the expiration date.
(b) The order shall require elimination of the violation by removal of the noncomplying prewrapped sandwich, in accordance with RSA 438:12.
(c) The order shall state a proposed administrative fine, not to exceed $1000 for each violation, pursuant to RSA 438:40 (IV).
(d) If the recipient of the order has no expiration date printed on the prewrapped sandwiches, the commissioner shall issue a stop-use order. The prewrapped sandwiches shall be allowed back on sale if the recipient of the stop-use order establishes within 4 hours an appropriate expiration date and prints it on the prewrapped sandwich(es).
(e) If the recipient of the order has kept the prewrapped sandwich(es) on the refrigerator shelf past its expiration date, the commissioner shall issue a stop-removal order, in accordance with RSA 438:13. The recipient of this order shall remove the offending sandwiches from his/her premises within 24 hours of its issuance.

Agr 1413.09 Appeal.

After an order has been issued by the commissioner the violator may appeal, pursuant to RSA 438:7 III, to the commissioner or to the superior court.

Uniform Open Dating Regulation
As adopted by The National Conference on Weights and Measures
(see Appendix C)

Mandatory compliance to the Uniform Open Dating Regulation.

NEW JERSEY

Regulations Establishing Definitions and Standards of Identity for Designated Fluid Milk Products, and Sanitary Requirements for the Production and Processing of Milk and Fluid Milk Products (Effective December 29, 1993)

SUBCHAPTER 10. DESIGNATED FLUID MILK PRODUCTS

8:21-10.1 Definitions and products standards

The following words and terms, when used in this chapter, shall have the following meanings unless the context clearly indicates otherwise. The following standards of identity conform to the Code of Federal Regulations for milk and cream (21 CFR 131).

8:21-10.12 Dating of milk and fluid milk products

(a) Fluid milk products as defined in N.J.S.A. 24:10-57.1 and all types and varieties of cottage and soft cheeses designated by the Department, intended for direct sale to consumers, shall be legibly marked with a "shelf-life expiration date." This date shall be determined and applied on the final consumer package or container by the initial processor or manufacturer. Prior to determining this date, each processor or manufacturer shall notify the Department of the intended date selected by him for each fluid milk product. All data and material used by the processor or manufacturer in his determination of this date shall be made available to the Commissioner upon request. If the data and material submitted does not, in the opinion of the Commissioner, justify the "shelf-life expiration date," the Commissioner shall prohibit the sale of the product until such time as satisfactory data is supplied or until a new "shelf-life expiration date," consistent with the data, is applied to the product.

(b) The packages or containers shall be marked with the legend "not to be sold after," or "sell by," or any other clearly understandable legend approved by the Department, followed by the "shelf-life expiration date". The designation of the month and date of the month after which the product shall not be sold may be numerical, such as "9-15" or "09 15" for September 15 or with the use of an abbreviation for the month such as "Sep 15" or "Se 15".

(c) The shelf-life expiration date "shall appear in a clear and legible manner and shall be placed on the part of the package or container most likely to be displayed, presented, or shown or examined under customary conditions of display for retail sale, and shall not interfere with the legibility of other mandatory labeling requirements of the product. However, cup containers that are labeled with the date on the bottom of the container shall have displayed on the cap or other conspicuous position information indicating the location of the date. The same provision applies for dates molded into plastic containers. Individual portion-pak containers not intended for direct resale to consumers shall be exempted, provided the bulk container which they are distributed is properly dated. Containers and packages of frozen cream and frozen desserts mixes not intended for resale to consumers shall also be exempted from the provisions of this regulation.

(d) No milk product referred to in this regulation shall be sold or offered for sale after 11:59 P.M. of the date appearing on the package or container. Products delivered prior to the "shelf-life expiration date" may be consumed on the premise beyond the date appearing thereon.

New Jersey Statutes Annotated
Dairy Products

24:10-57.23. Containers for milk and fluid milk products; dating and labeling; shelf life expiration date; shelf-life tests

Containers of milk, certified milk, Vitamin D milk, homogenized milk, low fat milk, protein fortified low fat milk, skim milk, protein fortified skim milk, nonfat milk, protein fortified nonfat milk, flavored milks and dairy drinks, buttermilk, cultured buttermilk, yogurt, eggnog, creams, half-and-half and all other fluid milk products designated by the department shall be marked with the name and address of the processor or the pasteurizing plant number as assigned by the department or the state of origin and the name and address of the distributor. All containers of fluid milk products, including those mentioned above, intended for sale to consumers, (except for those products which are sterilized and packaged in hermetically sealed containers), shall be marked with a legend "NOT TO BE SOLD AFTER", or "SELL BY", or any other clearly understandable legend approved by the department, followed or accompanied by the first three letters of the month where possible, but in no instance less than two letters, or numerical designation approved by the department to designate the month and the day of the month which shall be a date established by the processor and which shall be based on consideration of wholesomeness and consumer palatability of the product. If two letters are used the letters MR shall mean MARCH and MY shall mean MAY; JN shall mean JUNE and JL shall mean JULY. No fluid milk product listed in this section shall be sold or offered for sale after 11:59 p.m. of the date appearing on the containers so marked.

The processor, prior to determining the date beyond which any such fluid milk product may not be sold or offered for sale, shall notify the department of the intended "shelf-life expiration date" selected by him for such fluid milk product intended for sale. All data and material used by the processor or manufacturer in his determination of this date shall be made available to the commissioner upon request. If the data and material submitted does not, in the opinion of the commissioner, justify the "shelf-life expiration date" the commissioner shall prohibit the sale of the product until such time as satisfactory data is supplied or until a new "shelf-life expiration date" consistent with the data is applied to the product.

The department shall periodically review the keeping quality of milk and milk products by scientific shelf-life tests, recognizing the different methods of pasteurization, processing and packaging, to determine that shelf-life expiration dates stated on the containers assure the consumer of acceptable quality milk and milk products when kept under normal storage conditions. Samples for shelf-life evaluation will be obtained at the processing plant, from delivery trucks or from retail outlets. The temperature of the sample at the time of collection shall be officially recorded by the collector. Nothing herein contained shall be construed to prohibit the department from taking special samples for analysis and making special tests in order to assure all milk and milk products comply with the minimum standards of freshness, quality and palatability. In the event the department determines a processor's or manufacturer's shelf-life for a given product is improper, the department shall immediately take such samples as are necessary for full and complete recheck of the shelf-life of the product. If the full and complete recheck confirms that the shelf-life of the product is improper, the department shall serve written notice on the processor or manufacturer and the processor or manufacturer immediately upon receipt of such notice shall alter the shelf-life expiration date of the product to comply with the department findings. Compliance shall be with the next processing of the product after receipt of such department notice. This rule does not apply to containers of fluid milk products which are not to be sold in the State of New Jersey.

L.1964, c. 62, § 23, eff. May 18, 1964. Amended by L.1967, c. 92, § 1, eff. June 8, 1967; L.1972, c. 52, § 3, eff. June 1, 1962, operative Dec. 1, 1972; L.1979, c. 330, eff. Jan. 21., 1980; L.1992, c. 151, § 1, eff. Nov. 24, 1992.

NEW MEXICO

www.nmdaweb.nmsu.edu

5. OPEN DATE LABELING FOR DAIRY PRODUCTS

a. Definitions: As used in this regulatory order:
   (1) "pull date" means the last day on which a product is to be sold or offered for sale for human consumption.
   (2) "cultured product" means the product resulting from the souring by lactic acid producing bacteria or similar culture of a milk product.
   (3) "sterilized product" means a product that has been heated to a temperature of two hundred twelve (212°) degrees Fahrenheit or higher for a length of time sufficient to kill all organisms present.
   (4) "hermetically sealed product" means a product packaged in such a way as to be impervious to air and gasses, thus preventing microbial spoilage.
   (5) "ultra-pasteurized" means a product thermally processed at two hundred and eighty (280°) degrees Fahrenheit or higher for at least two (2) seconds followed by aseptic packaging so as to produce a product which has an extended shelf life under refrigerated conditions.
   (6) "dairy dessert mixes" means a product in fluid form intended to be used for dessert purposes and shall include, but not be limited to ice milk mix, ice cream mix, mellorine mix, yogurt mix, eggnog, or other product of similar nature.

b. Labeling Requirements for Dating: Superseded by NMDA Rule No. 93-1.

c. Dating Expression: The pull date shall be expressed numerically or by the first three letters of the month followed by the numeral or numerals constituting the appropriate calendar date. For example: December 1 may be expressed "Dec 1" or "12 01." A space, slash, or dash shall appear between the month and day.

d. Dating Location: The pull date shall appear in such a location on the container to be easily located and seen by the purchaser. The method of dating shall not interfere with legibility of other mandatory labeling requirements of the container.

e. Exempt Products: Those products sold directly to the consumer by a processor or producer distributor and all cultured, sterilized, ultra-pasteurized, or hermetically sealed products and dairy dessert mixes shall be exempt from pull date requirements. At the option of the processor, exempt products may be labeled with a pull date determined by the processor.

f. Compliance: Dairy products required to be labeled with a pull date, and those dairy products labeled with an optional pull date, except frozen, dried, condensed, or evaporated products may not be sold or offered for sale for human consumption by any person after the pull date.

g. Emergency Waiver: Any or all of the provisions of this regulation may be temporarily waived by the Director when he determines, in his discretion, that an emergency warrants such waiver.

NEW YORK

New York City Health Code
Article III Milk and Milk Products

111.33 Standards generally; time of delivery

No person shall possess, store, offer for sale, sell, give away or distribute milk, low sodium milk, low fat milk, skimmed milk, modified milk, cream or half and half after the expiration date indicated on the label required pursuant to Section 111.61. No person shall possess, store, offer for sale, sell, give away or distribute any such product the label of which bears an expiration date beyond the period specified in this section. The expiration date shall not be more than 96 hours after six a.m. following pasteurization. However, in the case of such milk and milk products which have been ultra-pasteurized pursuant to subsection (d) of Section 111.25, the expiration date shall not be more than fifteen (15) calendar days after six a.m. following ultra-pasteurization. This section does not apply to cans of milk or cream to be used for manufacturing purposes or to milk or milk products which are not to be sold in the City of New York.

NOTES: This section was originally derived from S.C. §156 Regs.51 and 155 (4). The new Health Code which became effective on October 1, 1959 contained a §111.33 which read the same as it did prior to the amendment hereafter referred to which was adopted on June 14, 1966. That section was repealed by resolution adopted on May 20, 1960 after §1400 of the Public Health Law was amended by the State Legislature so as to prohibit local boards of health from adopting and enforcing milk dating regulations. When that section was again amended by the State Legislature during the 1962 session by removing such prohibition this section was reenacted by resolution adopted on March 6, 1962.

The section was amended by resolution adopted on June 14, 1966, which changed the dating requirement in relation to the period beyond which milk and other milk products may not be possessed or sold. Formerly such period was 54 hours after the day of distribution in the case of milk and 72 hours after distribution in the case of cream and half and half. Under the section as amended such period is 66 hours from the day of pasteurization and the former distinction in this respect between milk, cream and half and half was eliminated. In addition, the dating requirement was made inapplicable to the products enzyme milk, flavored milk and flavored drink.

This section was further amended by resolution adopted on January 16, 1975 to add a dating requirement for ultra-pasteurized milk and milk products for which Federal standards of identity and quality were recently promulgated (21 CFR Part 18, revised September 28, 1973).

This section was further amended by resolution adopted on November 16, 1978 to extend the period from 66 hours to 96 hours from the day of pasteurization in which milk and milk products may be possessed or sold.

Section 111.25 Standards generally; pasteurization or ultra-pasteurization of milk and milk products required

Section 111.61 Labeling of containers

NORTH CAROLINA

North Carolina Administrative Code
Agriculture—Food and Drug Protection Division

Section .0500—Smoked and Smoke-Flavored Fish
.0502 Definitions

The following definitions apply:

(1) "Smoked fish" means any fish that is prepared by treating it with salt (sodium chloride) and then subjecting it to the direct action of smoke from burning wood, sawdust, or similar material.

(2) "Smoked-flavored fish" means any fish that is prepared by treating it with salt (sodium chloride) and then imparting to it the flavor of smoke by other than the direct action of smoke. This Paragraph does not alter the labeling requirements.

(3) "Hot process smoked or hot-process smoke-flavored fish" means the finished food prepared by subjecting forms of smoked fish to heat.

History Note: Statutory Authority G.S. 106-139; 106-267; 106-267.2; Eff. January 1, 1985

.0507 Processes and Controls

(e) Packing

(5) Permanently legible code marks shall be placed on the outer layer of every finished product package and master carton. Such marks shall identify at least the plant where packed, the date of packing, and the oven load. Records shall be so maintained as to provide positive identification;

History Note: Statutory Authority G.S. 106-139; 106-267;106-267; Eff. January 1, 1985.

NORTH DAKOTA

Voluntary

OHIO

Ohio Department of Agriculture
Division of Food, Dairies and Drugs
Sale Date on Perishable Food Products
Regulations 901:3-57-01 thru 901:3-57-08
Effective January 1, 1977

Regulation 901:3-57-01 Definitions

(A) "Quality Assurance Period" means the period of time following the completion of normal manufacturing, processing, and packaging procedures during which a food product subjected to normal conditions of exposure will maintain conformity with all of the characteristics normally associated with the food product and will provide the benefits for which the food product is normally purchased. Food product characteristics include, but are not limited to, taste, texture, smell, nutritional value, and reaction value with other food products if used as an ingredient with other food products.
(B) "Sale Date" means the date by which the manufacturer, processor, or packager of the packaged food product recommends that the food product be sold for consumption based on the food product's quality assurance period.
(C) "Packaged Perishable Food Product" means perishable food products packaged in consumer packages.
(D) "Perishable Food Product" means a food product, determined by the manufacturer to have a quality assurance period of thirty (30) days or less.
(E) The term "Commodity in package form" shall be construed to mean a commodity put up or packaged in any manner in advance of sale in units suitable for retail sale.
(F) A "consumer package" shall be construed to mean a commodity in package form that is customarily produced or distributed for sale through retail sales agencies or instrumentalities for consumption by individuals or use by individuals for the purposes of personal care or in the performance of services ordinarily rendered in or about the household or in connection with personal possessions.
(G) A "nonconsumer package" shall be construed to mean any commodity in package form other than a consumer package, and particularly a package intended solely for industrial or institutional use or for wholesale distribution.

Regulation 901:3-57-02 Food Products Considered Perishable

The manufacturer, processor or packager shall determine which food products manufactured, processed or packaged shall be considered perishable.

Regulation 901:3-57-03 Establishing Sale Dates for Perishable Foods

(A) The manufacturer, processor or packager shall establish the sale date for perishable food products.
(B) Sale dates established by the original manufacturer, processor or packager shall not be increased by a repacker unless the product has been reprocessed in a manner which would enhance or lengthen the quality assurance period.

Regulation 901:3-57-04 Sale Date—Information and Location

(A) No person shall knowingly sell a perishable food product in consumer size packages unless the package is clearly marked with a sale date by the packager. No other similarly stated date may appear on the package. However, nothing in this regulation shall prohibit a coded date, such as a Julian date, which may be required by Federal Law.
(B) The sale date markings shall be legible to the average consumer under normal conditions of purchase.
(C) Label declarations—the sale date may be indicated by appropriate descriptive terms such as, "Sell before," "sell by," etc. Terminology used to describe the sale date of a perishable food product shall be easily understood by the average consumer.
(D) Method of dating—The combination of words, numbers and/or abbreviations used to designate the sale date shall be easily understood by the average consumer. Words, numbers or abbreviations such as Nov. 23, 11-23, November 23, etc. are acceptable.

Bakery products with a sale date of less than seven days from the date of packaging may be dated with the days of the week or abbreviations of same in lieu of the foregoing requirements, as follows:

Sunday	SU, SUN	Thursday	TH, THU, THUR
Monday	MO, MON	Friday	FR, FRI
Tuesday	TU, TUES	Saturday	SA, SAT
Wednesday	WE, WED		

(E) Location of "Sale Date" information—"Sale Date" information shall be located on the package in a manner traditionally established by the trade and recognized by the consumer.

Regulation 901:3-57-05 Duties of Director

The Director is authorized to investigate complaints, to determine whether the sale date for food products as determined by the manufacturer, processor, or packager, is false or misleading to consumers. If the Director finds, upon reasonable cause, that the sale date as determined by the manufacturer, processor, or packager, is false or misleading to the consumer, the Director after reasonable notice and hearing, in accordance with Chapter 119 of the Revised Code, shall establish a sale date for said product.

Regulation 901:3-57-06 Exemptions

The provisions of this regulation do not apply to:
(A) Fresh fruits or vegetables
(B) Meat and meat products including poultry and poultry meat products whether packaged or unpackaged.
(C) Packaged perishable food products when sold or offered for sale at a place of business where less than one hundred thousand dollars ($100,000) of all products were sold during the preceding year.
(D) Food products in nonconsumer packages.

Regulation 901:3-57-07 Preemption Clause

To ensure that a uniform system of determining the useful product life of perishable food products for sale within the state is established, persons complying with this section and the rules established pursuant thereto are exempt from any local ordinances or rules pertaining to the quality assurance period of food products or the manner in which the quality assurance period and perishability of food products are to be disclosed.

Regulation 901:3-57-08 Penalties

Whoever violates section 3715.171 of the Ohio Revised Code or regulations adopted thereunder is guilty of a minor misdemeanor.

OKLAHOMA

Voluntary

OREGON

Title 49 Food and Other Commodities
(Purity; Sanitation; Grades; Standards; Labels; Weights and Measures)
Chapter 616
1995 Edition

OPEN DATE LABELING

616.800 Short title. ORS 616.800 to 616.835 and 616.994 may be cited as the Open Date Labeling Law. [1973 c. 173 §2]

616.805 Definitions for ORS 616.800 to 616.835 and 616.994. As used in ORS 616.800 to 616.835 and 616.994, unless the context requires otherwise:

(1) "Food" means any substance used or intended to be used for human consumption as food, drink or condiment.
(2) "Open date" means a date clearly visible to retail consumers showing the pull date, packing date or other date described in ORS 616.835 (2).
(3) "Packing date" means the date specifying the time a perishable food was packaged in its final form for sale to the consumer.
(4) "Perishable food" means any food that may spoil or otherwise become unfit for human consumption because of its nature, type or physical condition. "Perishable food" includes, but is not limited to, fresh or processed meats, poultry, seafood, dairy products, bakery products, eggs in the shell, and foods that have been packaged or refrigerated. ORS 616.800 to 616.835 and 616.994 shall not apply to fresh fruits or vegetables or to foods that have been canned or frozen.
(5) "Pull date" means, whichever is earlier, the date specifying the time:
(a) The perishable food manufacturer, processor or packager recommends that a perishable food should be removed from retail sale, allowing the consumer time for normal home consumption or use under proper care and storage conditions; or
(b) A perishable food should no longer be offered for sale or sold as fresh. A perishable food shall be considered fresh only so long as significant changes in appearance, taste, odor, nutritional value, or other indicia of quality or fitness for human consumption have not taken place or are not likely to have taken place under generally accepted food handling practices for that particular food. [1973 c.173 §3]

616.810 Exemption for alcoholic beverages. ORS 616.800 to 616.835 and 616.994 do not apply to alcoholic beverages. [1973 c.173 §9]

616.815 Open date labeling required for packaged perishable food sold at retail. No person shall sell or offer for sale at retail any packaged perishable food unless the package bears a clearly marked, printed or stamped label showing the open date for the perishable food in the package. Such label shall be so designed and placed as to be clearly visible to the consumer. [1973 c.173 §4]

616.820 Label required to be affixed to package not later than time of delivery to retail seller.

(1) The perishable food manufacturer, processor or packager shall affix, print or stamp the label required by ORS 616.815 to the perishable food retail package and to all closed shipping cartons, containers or wrappers of such perishable food packages not later than the time of delivery of the perishable food packages to the retail seller.
(2) No perishable food manufacturer, processor or packager shall fail to comply with subsection (1) of this section. [1973 c.173 §5]

616.825 Sale of perishable food after expiration of pull date prohibited; exceptions; time for removal of packages with expired pull dates.

(1) No person shall sell or offer for sale at retail any packaged perishable food after the expiration of the open pull date appearing on the label of the package or container unless:
(a) The package has been separated from packages of perishable food with open pull dates that have not expired;
(b) Each such package or group of packages is clearly identified in retail display as having an expired open pull date; and
(c) The food is fit for human consumption according to applicable state and federal law.
(2) Notwithstanding the provisions of this section, a vendor shall be allowed the first eight business hours after the expiration of the open pull date within which to remove all packages with an expired pull date. [1973 c.173 §6]

616.830 Altering labels or using nonconforming labels prohibited. No person shall:

(1) Alter, deface or remove the open date from any perishable food retail or shipping package carton, container or wrapper.
(2) Label any perishable food retail or shipping package carton, container or wrapper in a manner that does not conform to the rules promulgated pursuant to ORS 616.835. [1973 c.173 §7]

616.835 Rulemaking authority. In accordance with any applicable provision of ORS 183.310 to 183.550, the department, in consultation with the industries affected, shall promulgate rules to carry out ORS 616.800 to 616.835 and 616.994. Such rules shall include, but are not limited to:

(1) Establishing which particular foods are subject to ORS 616.800 to 616.835 and 616.994.

(2) Establishing which one or more of the following types of open date is to be used for particular groups or classes of perishable foods:
(a) The packing date.
(b) The pull date.
(c) The date on which fowl, including chickens, fryers, turkeys, ducks, geese and other domesticated birds, are killed or slaughtered to be processed into perishable food.
(3) Specifying the size, content and form of the labeling information required by ORS 616.800 to 616.835 and 616.994.
(4) Exempting from the operation of ORS 616.800 to 616.835 and 616.994 those perishable foods for which open date labeling would be:
(a) Impractical or not meaningful because of the size of the package or the nature of the perishable food;
(b) Possibly unconstitutional as interference with the free movement of goods in interstate commerce. [1973 c.173 §8]

PENNSYLVANIA

Commonwealth of Pennsylvania
Pennsylvania Code
Title 7. Agriculture
Part III. Bureau of Foods and Chemistry
Chapter 59. Milk Sanitation and Standards

§ 59.22. Milk dating.
(a) The cap or nonglass container of pasteurized milk held in stores, restaurants, schools or similar food establishments for resale shall be conspicuously and legibly marked in a contrasting color with the designation of the month and day of the month after which the product may not be sold or offered for sale. The designation may be numerical—such as "8-15"—or with the use of an abbreviation for the month, such as "AUG 15" or "AU 15." The words "Sell by" or "Not to be sold after" shall precede the designation of the date, or the statement "Not to be sold after the date stamped above" shall appear legibly on the container. This designation of the date may not exceed 14 days beginning after midnight on the day on which the dairy products were pasteurized. Stores, as used in this subsection, include mercantile establishments which offer milk for sale except on premises where processed.
(b) The date shall be separate and distinct from any other number, letter or intervening material on the cap or nonglass container.
(c) Pasteurized milk may not be sold after the date designated on the container.

(d) Sterile, ultra-pasteurized and cultured dairy products are exempt from the dating requirements.

AUTHORITY: The provisions of this § 59.22 amended under the act of July 2, 1935 (P.L. 589, No. 210) (31 P.S. §§ 645—660g).

SOURCE: The provisions of this § 59.22 amended September 4, 1987, effective September 5, 1988, 17 Pa.B. 3599; amended August 9, 1991, effective August 10, 1991, 21 Pa.B. 3508. Immediately preceding text appears at serial page (160963).

CROSS REFERENCES: This section cited in 7 Pa. Code § 59.21 (relating to general labeling requirements); and 7 Pa. Code § 59.32 (relating to sampling and examination).

RHODE ISLAND

Voluntary

SOUTH CAROLINA

Voluntary

SOUTH DAKOTA

South Dakota Department of Agriculture
Law 39-11
Rules and Regulations
Article 12:26
Pertaining to Eggs and Egg Products

Chapter 12:26:10 Labeling and Sales Requirements

12:26:10:01. **Identification and sale of graded eggs.** A person exposing or offering graded eggs for sale to consumers shall comply with the following requirements for identification and sale:

(3) The expiration date shall be legibly stamped on the outside of each carton. Eggs may be offered for sale for no more than 30 days after the date of pack. The expiration date using the day and the three-letter abbreviation for the month shall be preceded by the letters "EXP";

SOURCE: 2 SDR 7, effective August 4, 1975; 12 SDR 128, 12 SDR 154, effective July 1, 1986; 17 SDR 122, effective February 24, 1991; 19 SDR 61, effective October 26, 1992.

GENERAL AUTHORITY: SDCL 39-11-7.

LAW IMPLEMENTED: SDCL 39-11-7.

TENNESSEE

Voluntary

TEXAS

Voluntary

UTAH

Voluntary

VERMONT

Voluntary

VIRGINIA

Regulations Governing Grade "A" Milk
2VAC5-490-40. Labeling.
http://leg1.state.va.us/000/reg/TOC02005.HTM#C0490

m. The "pull date" which shall not interfere with the legibility of other labeling required for the milk or milk product and shall be expressed by: the first three letters in the name of the month, followed by or preceded by the numeral or numerals constituting the calendar date after which the product shall not be sold or expressed numerically by the number of the month followed by the number of the day. For example, June 1 shall be expressed "JUN 1," "1 JUN," "06 01," or "06-01." Nothing in this chapter pertaining to pull dates shall apply to grade A pasteurized milk and grade A pasteurized milk products bottled in glass containers for home delivery;
n. The grade A permit holder who operates a milk plant and offers for sale milk or milk product within the Commonwealth shall file and certify with the State Regulatory Authority the maximum number of days after manufacturing or processing the grade A permit holder's milk or milk products which will be used to determine the "pull date." The grade A permit holder shall establish a "pull date" that under normal storage the milk or milk product meets for a minimum of 96 hours after the "pull date," standards set by this chapter;
o. No person may sell or offer for sale any grade A pasteurized milk, grade A pasteurized milk product, or milk product in a package that does not bear the "pull date;" and
p. Nothing in this chapter shall apply to containers of grade A pasteurized milk, grade A milk products, or milk products which are not to be sold in the Commonwealth.

WASHINGTON

Chapter 69.04 RCW
Intrastate Commerce in Food, Drugs, and Cosmetics
(Formerly: Food, drug, and cosmetic act)

RCW 69.04.900 Perishable packaged food—Pull date labeling—Definitions. For the purpose of RCW 69.04.900 through 69.04.920:

(1) "Perishable packaged food goods" means and includes all foods and beverages, except alcoholic beverages, frozen foods, fresh meat, poultry and fish and a raw agricultural commodity as defined in this chapter, intended for human consumption which are canned, bottled, or packaged other than at the time and point of retail sale, which have a high risk of spoilage within a period of thirty days, and as determined by the director of the department of agriculture by rule and regulation to be perishable.
(2) "Pull date" means the latest date a packaged food product shall be offered for sale to the public.
(3) "Shelf life" means the length of time during which a packaged food product will retain its safe consumption quality if stored under proper temperature conditions.
(4) "Fish" as used in subsection (1) of this section shall mean any water breathing animals, including, but not limited to, shellfish such as lobster, clams, crab, or other mollusca which are prepared, processed, sold, or intended or offered for sale. [1974 ex.s. c 57 § 1; 1973 1st ex.s. c 112 § 1.]

RCW 69.04.905 Perishable packaged food—Pull date labeling—Required. All perishable packaged food goods with a projected shelf life of thirty days or less, which are offered for sale to the public after January 1, 1974 shall state on the package the pull date. The pull date must be stated in day, and month and be in a style and format that is readily decipherable by consumers: PROVIDED, That the director of the department of agriculture may exclude the monthly requirement on the pull date for perishable packaged food goods which have a shelf life of seven days or less. No perishable packaged food goods shall be offered for sale after the pull date, except as provided in RCW 69.04.910. [1974 ex.s. c 57 § 2; 1973 1st ex.s c 112 § 2.]

RCW 69.04.910 Perishable packaged food—Pull date labeling—Selling or trading goods beyond pull date—Repackaging to substitute for original date—Exception. No person shall sell, trade or barter any perishable packaged food goods beyond the pull date appearing thereon, nor shall any person rewrap or repackage any packaged perishable food goods with the intention of placing a pull date thereon which is different from the original: PROVIDED,

HOWEVER, That those packaged perishable food goods whose pull dates have expired may be sold if they are still wholesome and are without danger to health, and are clearly identified as having passed the pull date. [1973 1st ex.s. c 112 § 3.]

RCW 69.04.915 Perishable packaged food—Pull date labeling—Storage—Rules and regulations. The director of the department of agriculture shall by rule and regulation establish uniform standards for pull date labeling, and optimum storage conditions of perishable packaged food goods. In addition to his other duties the director, in consultation with the secretary of the department of health where appropriate, may promulgate such other rules and regulations as may be necessary to carry out the purposes of RCW 69.04.900 through 69.04.920. [1989 1st ex.s. c 9 § 225; 1973 1st ex.s. c 112 § 4.]

Effective date—Severability—1989 1st ex.s. c 9: See RCW 43.70.910 and 43.70.920.

RCW 69.04.920 Perishable package food—Pull date labeling—Penalties. Any person convicted of a violation of RCW 69.04.905or 69.04.910 shall be punishable by a fine not to exceed five hundred dollars. [1973 1st ex.s. c 112 § 5.]

WEST VIRGINIA

West Virginia Code (1996)

§ 47-1-9. Requirements for open dating.

The uniform open dating regulation as adopted by the national conference on weights and measures and published in national institute of standards and technology handbook 130, "Uniform Laws and Regulations" and supplements thereto and revisions thereof, shall apply to open dating in the state, except insofar as modified or rejected by legislative rule. (1994, c. 178.)

(See Appendix C)

WISCONSIN

Agriculture, Trade & Consumer Protection
Chapter ATCP 70 Food Processing Plants

ATCP 70.22 Labeling and sale of smoked fish.

(1) Every food package containing smoked fish shall be clearly and conspicuously labeled, on the principal display panel of that package, with all of the following information:
  - (f) The processing date of the smoked fish.
  - (g) The last date on which the smoked fish may be sold at retail, which shall be not more than 17 days after the processing date under par. (f). This paragraph does not apply to any of the following:
    - (1) Smoked fish which are distributed and sold in a frozen state.
    - (2) Smoked fish which are clearly labeled to indicate that they must be cooked by the consumer prior to consumption.
    - (3) Smoked fish which have more than 10% salt content.
    - (4) Smoked fish which are specifically exempted by the department in writing.
  - (2) Smoked fish shall not be sold more than 17 days after the date of processing unless the smoked fish are exempt under sub. (1)(g).
  - (3) Smoked fish processed on different dates may not be commingled in the same container, either at the processing plant or while the fish is being stored, distributed or offered for sale at the wholesale or retail.
  - (4) No person may misrepresent a smoked fish processing date, or sell or distribute smoked fish under any processing date other than the original processing date stated by the processor.
  - (5) Food consisting of or containing smoked fish shall be immediately removed from sale, and shall be destroyed or treated to render it unattractive and unfit for human consumption, if any of the following occurs:
    - (a) The food package is not labeled with a processing date.
    - (b) The food package is not labeled with a final date of sale unless its smoked fish contents are exempt under sub. (1)(g).
    - (c) The food is not sold at retail within 21 days after the date on which its smoked fish contents were smoked, unless the smoked fish are exempt under sub. (1)(g).
    - (d) The food is held at a temperature above 38°F (3.4°C) at any time prior to retail sale. This paragraph does not apply to a food which the department specifically exempts in writing because it is not a potentially hazardous food.

WYOMING

Voluntary

APPENDIX E

European Union

Council Directive of 18 December 1978 on the approximation of the laws of the Member States relating to the labelling, presentation and advertising of foodstuffs (79/112/EEC)

as adopted by
Food Law in the EU Version 1:2. Published by Chapman & Hall

Note: This Directive is amended by:
Directive 97/4/EEC of the European Parliament and of the Council

ARTICLE 1

1. This Directive concerns the labelling of foodstuffs to be delivered as such to the ultimate consumer and certain aspects relating to the presentation and advertising thereof.
2. This Directive shall apply also to foodstuffs intended for supply to restaurants, hospitals, canteens and other similar mass caterers (hereinafter referred to as "mass caterers").
3. For the purpose of this Directive,
   (a) "Labelling" shall mean any words, particulars, trade marks, brand name, pictorial matter or symbol relating to a foodstuff and placed on any packaging, document, notice, label, ring or collar accompanying or referring to such foodstuff;
   (b) "Pre-packaged foodstuff" shall mean any single item for presentation as such to the ultimate consumer and to mass caterers, consisting of a foodstuff and the packaging into which it was put before being offered for sale, whether such packaging encloses the foodstuff completely or only partially, but in any case in such a way that the contents cannot be altered without opening or changing the packaging.

ARTICLE 2

1. The labelling and methods used must not:
   (a) Be such as could mislead the purchaser to a material degree, particularly:
      (i) As to the characteristics of the foodstuff and, in particular, as to its nature, identity, properties, composition, quantity, durability, origin or provenance, method of manufacture or production,
      (ii) By attributing to the foodstuff effects or properties which it does not possess,

(iii) By suggesting that the foodstuff possesses special characteristics when in fact all similar foodstuffs possess such characteristics;

(b) Subject to Community provisions applicable to natural mineral waters and foodstuffs for particular nutritional uses, attribute to any foodstuff the property of preventing, treating or curing a human disease, or refer to such properties.

2. The Council, in accordance with the procedure laid down in Article 100 of the Treaty, shall draw up a non-exhaustive list of the claims within the meaning of paragraph 1, the use of which must at all events be prohibited or restricted.

3. The prohibitions or restrictions referred to in paragraphs 1 and 2 shall also apply to:

(a) The presentation of foodstuffs, in particular their shape, appearance or packaging, the packaging materials used, the way in which they are arranged and the setting in which they are displayed;

(b) Advertising.

ARTICLE 3

1. In accordance with Articles 4 to 14 and subject to the exceptions contained therein, indication of the following particulars alone shall be compulsory on the labelling of foodstuffs:

(1) The name under which the product is sold,

(2) The list of ingredients,

(3) In the case of prepackaged foodstuffs, the net quantity,

(4) The date of minimum durability or, in the case of foodstuffs which, from the microbiological point of view, are highly perishable, the "use by" date,

(5) Any special storage conditions or conditions of use,

(6) The name or business name and address of the manufacturer or packager, or of a seller established within the Community.

However, the Member States shall be authorized, in respect of butter produced in their territory, to require only an indication of the manufacturer, packager or seller.

Without prejudice to the notification provided for in Article 22, Member States shall inform the Commission and the other Member States of any measure taken pursuant to this paragraph,

(7) Particulars of the place of origin or provenance in the cases where failure to give such particulars might mislead the consumer to a material degree as to the true origin or provenance of the foodstuff,

(8) Instructions for use when it would be impossible to make appropriate use of the foodstuff in the absence of such instructions,

(9) With respect to beverages containing more than 1,2% by volume of alcohol, the actual alcoholic strength by volume.

2. Notwithstanding the previous paragraph, Member States may retain national provisions which require indication of the factory or packaging centre, in respect of home production.
3. The provisions of this Article shall be without prejudice to more precise or more extensive provisions regarding weights and measures.

ARTICLE 4

1. Community provisions applicable to specified foodstuffs and not to foodstuffs in general may provide for derogations, in exceptional cases, from the requirements laid down in Article 3 (1), points 2 and 4, provided that this does not result in the purchaser being inadequately informed.
2. Community provisions applicable to specified foodstuffs and not to foodstuffs in general may provide that other particulars in addition to those listed in Article 3 must appear on the labelling.

   Where there are no Community provisions, Member States may make provision for such particulars in accordance with the procedure laid down in Article 16.

3. The Community provisions referred to in paragraphs 1 and 2 shall be adopted in accordance with the procedure laid down in Article 17.

ARTICLE 5

1. The name under which a foodstuff is sold shall be the name laid down by whatever laws, regulations or administrative provisions apply to the foodstuff in question or, in the absence of any such name, the name customary in the Member State where the product is sold to the ultimate consumer and to mass caterers, or a description of the foodstuff and, if necessary, of its use, that is sufficiently precise to inform the purchaser of its true nature and to enable it to be distinguished from products with which it could be confused.
2. No trade mark, brand name or fancy name may be substituted for the name under which the product is sold.
3. The name under which the product is sold shall include or be accompanied by particulars as to the physical condition of the foodstuff or the specific treatment which it has undergone (e.g., powdered, freeze-dried, deep-frozen, concentrated, smoked) in all cases where omission of such information could create confusion in the mind of the purchaser.

Any foodstuff which has been treated with ionizing radiation must bear one of the following indications:

In English:	'irradiated' or 'treated with ionizing radiation'
In Spanish:	'irradiado' or 'tratado con radiacíon ionizante'
In Danish:	'bestrålet/...' or 'strålekonserveret' or 'behandlet med ioniserende stråling' or 'konserveret med ioniserende stråling'
In German:	'bestrahlt' or 'mit ionisierenden Strahlen behandelt'
In Greek:	' ...'
In French:	'traité par rayonnements ionisants' or 'traité par ionisation'
In Italian:	'irradiato' or 'trattato con radiazioni ionizzanti'
In Dutch:	'doorstraald' or 'door bestraling behandeld' or 'met ioniserende stralen behandeld'
In Portuguese:	'irradiado' or 'tratado por irradiações' or 'tratado por radiação ionizante'
In Finnish:	'säteilytetty, käsitelty inoisoivalla säteilylla'
In Swedish:	'bestrålad, behandlad med joniserande strålning'

ARTICLE 6

1. Ingredients shall be listed in accordance with this Article and the Annexes.
2. Ingredients need not be listed in the case of:
   (a) -- fresh fruit and vegetables, including potatoes, which have not been peeled, cut or similarly treated,
   -- carbonated water, the description of which indicates that it has been carbonated,
   -- fermentation vinegars derived exclusively from a single basic product, provided that no other ingredient has been added;
   (b) --cheese,
   --butter,
   --fermented milk and cream,
   Provided that no ingredient has been added other than lactic products, enzymes and micro-organism cultures essential to manufacture, or the salt needed for the manufacture of cheese other than fresh cheese and processed cheese;
   (c) Products consisting of a single ingredient.
3. In the case of beverages containing more than 1,2% by volume of alcohol, the Council, acting on a proposal from the commission, shall, before the expiry of a period of four years following notification of this directive, determine the rules for labelling ingredients.

4. (a) "Ingredient" shall mean any substance, including additives, used in the manufacture or preparation of a foodstuff and still present in the finished product, even if in altered form.
   (b) Where an ingredient of the foodstuff is itself the product of several ingredients, the latter shall be regarded as ingredients of the foodstuff in question.
   (c) The following shall not be regarded as ingredients:
      (i) The constituents of an ingredient which have been temporarily separated during the manufacturing process and later reintroduced but not in excess of their original proportions;
      (ii) - additives:
         -- whose presence in a given foodstuff is solely due to the fact that they were contained in one or more ingredients of that foodstuff, provided that they serve no technological function in the finished product,
         -- which are used as processing aids;
      - substances used in the quantities strictly necessary as solvents or media for additives or flavouring.
   (d) In certain cases decisions may be taken in accordance with the procedure laid down in Article 17 as to whether the conditions described in (c) (ii) are satisfied.
5. (a) The list of ingredients shall include all the ingredients of the foodstuff, in descending order of weight, as recorded at the time of their use in the manufacture of the foodstuff. It shall appear preceded by a suitable heading which includes the word "ingredients".

However:

-- Added water and volatile products shall be listed in order of their weight in the finished product; the amount of water added as an ingredient in a foodstuff shall be calculated by deducting from the total amount of the finished product the total amount of the other ingredients used. This amount need not be taken into consideration if it does not exceed 5% by weight of the finished product;
-- Ingredients used in concentrated or dehydrated form and reconstituted at the time of manufacture may be listed in order of weight as recorded before their concentration or dehydration;
-- In the case of concentrated or dehydrated foods which are intended to be reconstituted by the addition of water, the ingredients may be listed in order of proportion in the reconstituted product provided that the list of ingredients is accompanied by an expression such as "ingredients of the reconstituted product", or "ingredients of the ready-to-use product";

-- In the case of mixtures of fruit or vegetables where no particular fruit or vegetable significantly predominates in proportion by weight, those ingredients may be listed in another order provided that that list of ingredients is accompanied by an expression such as "in variable proportion";
-- In the case of mixtures of spices or herbs, where none significantly predominates in proportion by weight, those ingredients may be listed in another order provided that that list of ingredients is accompanied by an expression such as "in variable proportion";

(b) Ingredients shall be designated by their specific name, where applicable, in accordance with the rules laid down in Article 5.

However:

-- Ingredients which belong to one of the categories listed in Annex I and are constituents of another foodstuff may be designated by the name of that category only. Alterations to the list of categories in Annex I may be effected in accordance with the procedure laid down in Article 17;
-- Ingredients belonging to one of the categories listed in Annex II must be designated by the name of that category, followed by their specific name or EEC number; if an ingredient belongs to more than one of the categories, the category appropriate to the principal function in the case of the foodstuff in question shall be indicated; amendments to this Annex based on advances in scientific and technical knowledge shall be adopted in accordance with the procedure laid down in Article 17;
-- Flavourings shall be designated in accordance with Annex III to this Directive. These provisions shall be adopted in accordance with the procedure laid down in Article 17;
-- The specific Community provisions governing the indication of treatment of an ingredient with ionizing radiation shall be adopted subsequently in accordance with Article 100a of the Treaty.

6. Community provisions or, where there are none, national provisions may lay down that the name under which a specific foodstuff is sold is to be accompanied by mention of a particular ingredient or ingredients. The procedure laid down in Article 16 shall apply to any such national provisions.

   The Community provisions referred to in this paragraph shall be adopted in accordance with the procedure laid down in Article 17.

7. In the case referred to in paragraph 4 (b), a compound ingredient may be included in the list of ingredients, under its own designation in so far as this is laid down by law or established by custom, in terms of its overall weight, provided that it is immediately followed by a list of its ingredients.

Such a list, however, shall not be compulsory:
-- where the compound ingredient constitutes less than 25% of the finished product; however, this exemption shall not apply in the case of additives, subject to the provisions of paragraph 4 (c),
-- where the compound ingredient is a foodstuff for which a list of ingredients is not required under Community rules.

8. Notwithstanding paragraph 5 (a), the water content need not be specified:
   (a) Where the water is used during the manufacturing process solely for the reconstitution of an ingredient used in concentrated or dehydrated form;
   (b) In the case of a liquid medium which is not normally consumed.

ARTICLE 7

1. Where the labelling of a foodstuff places emphasis on the presence or low content of one or more ingredients which are essential to the specific properties of the foodstuff, or where the description of the foodstuff has the same effect, the minimum or maximum percentage, as the case may be, used in the manufacture thereof shall be stated.

   This information shall appear either immediately next to the name under which the foodstuff is sold or in the list of ingredients in connection with the ingredient in question.

   In accordance with the procedure laid down in Article 17, it may be decided that, in the case of certain ingredients, the percentage referred to in this paragraph shall be expressed in absolute terms.
2. Paragraph 1 shall not apply:
   (a) In the case of labelling which is intended to characterize a foodstuff in accordance with Article 5 (1) or which is required under community provisions or, where there are none, under national provisions applicable to certain foodstuffs;
   (b) In the case of ingredients used in small quantities only as flavourings.
3. Community provisions or, where there are none, national provisions may stipulate for certain foodstuffs, as well as in the case referred to in paragraph 2 (a), that quantities of certain ingredients must be indicated either in absolute terms or as percentages and that, where appropriate, mention should be made of any alteration in the quantities of these ingredients.

   The procedure laid down in Article 16 shall apply to any such national provisions.

   The Community provisions referred to in this paragraph shall be adopted in accordance with the procedure laid down in Article 17.

ARTICLE 8

1. The net quantity of prepackage foodstuffs shall be expressed:
   -- in units of volume in the case of liquids,
   -- in units of mass in the case of other products,
   using the litre, centilitre, millilitre, kilogram or gram, as appropriate.

   Community provisions or, where there are none, national provisions applicable to certain specified foodstuffs may derogate from this rule.

   The procedure laid down in Article 16 shall apply to any such national provisions.
2. (a) Where the indication of a certain type of quantity (e.g. nominal quantity, minimum quantity, average quantity) is required by community provisions or, where there are none, by national provisions, this quantity shall be regarded as the net quantity for the purposes of this directive.

   Without prejudice to the notification provided for in Article 22, Member States shall inform the Commission and the other Member States of any measure taken pursuant to this point.

   (b) Community provisions or, where there are none, national provisions may, for certain specified foodstuffs classified by quantity in categories, require other indications of quantity.

   The procedure laid down in Article 16 shall apply to any such national provisions.

   (c) Where a prepackaged item consists of two or more individual prepackaged items containing the same quantity of the same product, the net quantity shall be indicated by mentioning the net quantity contained in each individual package and the total number of such packages. Indication of these particulars shall not, however, be compulsory where the total number of individual packages can be clearly seen and easily counted from the outside and where at least one indication of the net quantity contained in each individual package can be clearly seen from the outside.

   (d) Where a prepackaged item consists of two or more individual packages which are not regarded as units of sale, the net quantity shall be given by indicating the total net quantity and the total number of individual packages. Community provisions or, where there are none, national provisions need not, in the case of certain foodstuffs, require indication of the total number of individual packages. Without prejudice to the notification provided for in Article 22, Member States shall inform the Commission and the other Member States of any measure taken pursuant to this point.

3. In the case of foodstuffs normally sold by number, Member States need not require indication of the net quantity provided that the number of items can clearly be seen and easily counted from the outside or, if not, is indicated on the labelling.

   Without prejudice to the notification provided for in Article 22, Member States shall inform the Commission and the other Member States of any measure taken pursuant to this paragraph.
4. Where a solid foodstuff is presented in a liquid medium, the drained net weight of the foodstuff shall also be indicated on the labelling.

   For the purposes of this paragraph, "liquid medium" shall mean the following products, possibly in mixtures and also where frozen or quick-frozen, provided that the liquid is merely an adjunct to the essential elements of that preparation and is thus not a decisive factor for the purchase: water, aqueous solutions of salts, brine; aqueous solutions of food acids, vinegar; aqueous solutions of sugar, aqueous solutions of other sweetening substances; fruit or vegetable juices in the case of fruit or vegetables.

   This list may be supplemented in accordance with the procedure laid down in Article 17.

   Methods of checking the drained net weight shall be determined in accordance with the procedure laid down in Article 17.
5. It shall not be compulsory to indicate the net quantity in the case of foodstuffs:
   (a) Which are subject to considerable losses in their volume or mass and which are sold by number or weighed in the presence of the purchaser;
   (b) The net quantity of which is less than 5 g or 5 ml; however this provision shall not apply to spices and herbs.

   Community provisions or, where there are none, national provisions applicable to specified foodstuffs may in exceptional cases lay down thresholds which are higher than 5 g or 5 ml provided that this does not result in the purchaser being inadequately informed.

   Without prejudice to the notification provided for in Article 22, Member States shall inform the Commission and the other Member States of any measure taken pursuant to this paragraph.
6. Until the end of the transitional period during which the use of the imperial units of measurement contained in Chapter D of the Annex to Directive 71/354/EEC of 18 October 1971 on the approximation of the laws of the Member States relating to units of measurements, as last amended by Directive 76/770/EEC, is authorized in the Community, Ireland and the United Kingdom may permit the quantity to be expressed only in imperial

units of measurement calculated on the basis of the following conversion rates:

-- 1 ml = 0,0352 fluid ounces,
-- 1 l = 1,760 pints or 0,220 gallons,
-- 1 g = 0,0353 ounces (avoirdupois),
-- 1 kg = 2,205 pounds.

7. The Community provisions referred to in paragraphs 1, 2 (b) and (d) and 5 shall be adopted in accordance with the procedure laid down in Article 17.

ARTICLE 9

1. The date of minimum durability of a foodstuff shall be the date until which the foodstuff retains its specific properties when properly stored. It shall be indicated in accordance with the provisions of this article.
2. The date shall be preceded by the words:
   -- "Best before..." when the date includes an indication of the day,
   -- "Best before end..." in other cases.
   "..."
3. The words referred to in paragraph 2 shall be accompanied by:
   -- either the date itself, or
   -- a reference to where the date is given on the labelling.
   If need be, these particulars shall be followed by a description of the storage conditions which must be observed if the product is to keep for the specified period.
4. The date shall consist of the day, month and year in uncoded chronological form.
   However, in the case of foodstuffs:
   -- which will not keep for more than three months, an indication of the day and the month will suffice,
   -- which will keep for more than three months but not more than 18 months, an indication of the month and year will suffice,
   -- which will keep for more than 18 months, an indication of the year will suffice.
   The manner of indicating the date may be specified according to the procedure laid down in Article 17.
5. In their own territories the Member States may, until 31 December 1992, permit the minimum durability period to be expressed otherwise than in terms of the date of minimum durability. Without prejudice to the notification provided for in Article 22, Member States shall notify the Commission and the other Member States of any measure taken under this paragraph.

6. Subject to Community provisions imposing other types of date indication, an indication of the durability date shall not be required for:
   -- fresh fruit and vegetables, including potatoes, which have not been peeled, cut or similarly treated. This derogation shall not apply to sprouting seeds and similar products such as legume sprouts,
   -- wines, liqueur wines, sparkling wines, aromatized wines and similar products obtained from fruits other than grapes, and beverages falling within CN codes 22.06 and manufactured from grapes or grape musts,
   -- beverages containing 10% or more by volume of alcohol,
   -- soft drinks, fruit juices, fruit nectars and alcoholic beverages in individual containers of more than five litres, intended for supply to mass caterers,
   -- bakers' or pastry cooks' wares which, given the nature of their content, are normally consumed within 24 hours of their manufacture,
   -- vinegar,
   -- cooking salt,
   -- solid sugar,
   -- confectionery products consisting almost solely of flavoured and/or coloured sugars,
   -- chewing gums and similar chewing products,
   -- individual portions of ice-cream.

ARTICLE 9a

1. In the case of foodstuffs which, from the microbiological point of view, are highly perishable and are therefore likely after a short period to constitute an immediate danger to human health, the date of minimum durability shall be replaced by the "use by" date.
2. The date shall be preceded by the words:

   In English: 'use by'
   In Spanish: 'fecha de caducidad'
   In Danish: 'sidste anvendelsesdato'
   In German: 'verbrauchen bis'
   In Greek: '...'
   In French: 'à consommer jusqu aú'
   In Italian: 'da consumare entro'
   In Dutch: 'te gebruiken tot'
   In Portuguese: 'a consumir até'
   In Finnish: 'viimeinen käyttöajankohta'
   In Swedish: 'sista förbrukningsdag'

   These words shall be accompanied by:
   -- either the date itself, or

-- a reference to where the date is given on the labelling.

These particulars shall be followed by a description of the storage conditions which must be observed.

3. The date shall consist of the day, the month and, possibly, the year, in that order and in uncoded form.
4. In some cases it may be decided by the procedure laid down in Article 17 whether the conditions laid down in paragraph 1 are fulfilled.

ARTICLE 10

1. The instructions for use of a foodstuff shall be indicated in such a way as to enable appropriate use to be made thereof.
2. Community provisions or, where there are none, national provisions may, in the case of certain foodstuffs, specify the way in which the instructions for use should be indicated.

   The procedure laid down in Article 16 shall apply to such national provisions.

   The Community provisions referred to in this paragraph shall be adopted in accordance with the procedure laid down in Article 17.

ARTICLE 10a

The rules concerning indication of the alcoholic strength by volume shall, in the case of products covered by tariff heading nos 22.04, be those laid down in the specific Community provisions applicable to such products.

In the case of other beverages containing more than 1,2% by volume of alcohol, these rules shall be laid down in accordance with the procedure provided for in Article 17.

ARTICLE 11

1. (a) When the foodstuffs are prepackaged, the particulars provided for in Articles 3 and 4 (2) shall appear on the prepackaging or on a label attached thereto.
   (b) Notwithstanding point (a) and without prejudice to Community provisions on nominal quantities, where prepackaged foodstuffs are:
      -- intended for the ultimate consumer but marketed at a stage prior to sale to the ultimate consumer and where sale to a mass caterer is not involved at that stage,
      -- intended for supply to mass caterers for preparation, processing, splitting or retail sale,

the particulars required under Articles 3 and 4 (2) need appear only on the commercial documents referring to the foodstuffs where it can be guaranteed that such documents, containing all the labelling information, either accompany the foodstuffs to which they refer or were sent before or at the same time as delivery.

(c) In the cases referred to in (b), the particulars referred to in Article 3 (1) (1), (4) and (6) and, where appropriate, that referred to in Article 9a, shall also appear on the external packaging in which the foodstuffs are presented for marketing.

2. These particulars shall be easy to understand and marked in a conspicuous place in such a way as to be easily visible, clearly legible and indelible.

   They shall not in any way be hidden, obscured or interrupted by other written or pictorial matter.

3. (a) The particulars listed in Article 3 (1), points 1, 3, 4 and 9 shall appear in the same field of vision. This requirement may be extended to the particulars provided for in Article 4 (2).
   (b) However, for glass bottles intended for re-use, upon which one of the particulars listed in point (a) is indelibly marked, this requirement shall not apply for a period of 10 years following notification of this Directive.
4. In the case of the glass bottles intended for re-use which are indelibly marked and which therefore bear no label, ring or collar and packaging or containers the largest surface of which has an area of less than 10 cm^2 only the particulars listed in Article 3 (1) (1), (3) and (4) need be given. In this case, paragraph 3 (a) shall not apply.
5. Member States may, until 31 December 1996, refrain from requiring the minimum durability date or the "use by" date to be mentioned in respect of bottles referred to in paragraph 4.
6. Ireland, the Netherlands and the United Kingdom may derogate from Article 3 (1) and paragraph 3 (a) of this Article in the case of milk and milk products put up in glass bottles intended for re-use.
7. The Member States shall inform the Commission of any measure taken pursuant to paragraphs 5 or 6.

ARTICLE 12

Where foodstuffs are offered for sale to the ultimate consumer or to mass caterers without prepackaging, or where foodstuffs are packaged on the sales premises at the consumer's request or prepackaged for direct sale, the Member States shall adopt detailed rules concerning the manner in which the particulars specified in Article 3 and Article 4 (2) are to be shown.

They may decide not to require the provision of all or some of these particulars, provided that the purchaser still receives sufficient information.

ARTICLE 13

This Directive shall not affect the provisions of national laws which, in the absence of Community provisions, impose less stringent requirements for the labelling of foodstuffs presented in fancy packaging such as figurines or souvenirs.

ARTICLE 14

Member States shall refrain from laying down requirements more detailed than those already contained in Articles 3 to 11 concerning the manner in which the particulars provided for in Article 3 and Article 4 (2) are to be shown.

The Member States shall, however, ensure that the sale of foodstuffs within their own territories is prohibited if the particulars provided in Article 3 and Article 4 (2) do not appear in a language easily understood by purchasers, unless other measures have been taken to ensure that the purchaser is informed. This provision shall not prevent such particulars from being indicated in various languages.

ARTICLE 15

1. Member States may not forbid trade in foodstuffs which comply with the rules laid down in this Directive by the application of non-harmonized national provisions governing the labelling and presentation of certain foodstuffs or of foodstuffs in general.
2. Paragraph 1 shall not apply to non-harmonized national provisions justified on grounds of:
   -- protection of public health,
   -- prevention of fraud, unless such provisions are liable to impede the application of the definitions and rules laid down by this directive,
   -- protection of industrial and commercial property rights, indications of provenance, registered designations of origin and prevention of unfair competition.

ARTICLE 16

Where reference is made to this Article, the following procedure shall apply:

1. When a Member State maintains the provisions of its national laws, it shall inform the Commission and the other Member States thereof within a period of two years after notification of this Directive;

2. Should a Member State deem it necessary to adopt new legislation, it shall notify the Commission and the other Member States of the measures envisaged and give the reasons justifying them. The Commission shall consult the Member States within the Standing Committee on Foodstuffs if it considers such consultation to be useful or if a Member State so requests.

   Member States may take such envisaged measures only three months after such notification and provided that the Commission's opinion is not negative.

   In the latter event, and before the expiry of the abovementioned period, the Commission shall initiate the procedure provided for in Article 17 in order to determine whether the envisaged measures may be implemented subject, if necessary, to the appropriate modifications.

ARTICLE 17

Where the procedure laid down in this Article is to be followed, the matter shall be referred to the Standing Committee on Foodstuffs (hereinafter called "the Committee") by its chairman, either on his own initiative or at the request of a representative of a Member State.

The representative of the Commission shall submit to the Committee a draft of the measures to be taken. The Committee shall deliver its opinion on the draft within a time limit which the chairman may lay down according to the urgency of the matter. The opinion shall be delivered by the majority laid down in Article 148 (2) of the Treaty in the case of Decisions which the Council is required to adopt on a proposal from the Commission. The votes of the representatives of the Member States within the Committee shall be weighted in the manner set out in that Article. The chairman shall not vote.

The Commission shall adopt the measures envisaged if they are in accordance with the opinion of the Committee.

If the measures envisaged are not in accordance with the opinion of the Committee, or if no opinion is delivered, the Commission shall, without delay, submit to the Council a proposal relating to the measures to be taken. The Council shall act by a qualified majority.

If, on the expiry of a period three months from the date of referral to the Council, the Council has not acted, the proposed measures shall be adopted by the Commission.

ARTICLE 18

...

ARTICLE 19

If temporary measures prove necessary to facilitate the application of this Directive, they shall be adopted in accordance with the procedure provided for in Article 17.

ARTICLE 20

This Directive shall not affect Community provisions relating to the labelling and presentation of certain foodstuffs already adopted at the time of its notification. Any amendments necessary to harmonize such provisions with the rules laid down in this Directive shall be decided in accordance with the procedure applicable to each of the provisions in question.

ARTICLE 21

This Directive shall not apply to products for export outside the Community.

ARTICLE 22

1. Member States shall make such amendments to their laws as may be necessary to comply with the provisions of this Directive and shall forthwith inform the Commission thereof; the laws thus amended shall be applied in such a way as to:
   -- permit trade in those products which comply with the provisions of this Directive no later than two years after its notification,
   -- prohibit trade in those products which do not comply with the provisions of this Directive four years after its notification.
2. However, Member States may:
   (a) In the case of certain foodstuffs, reduce the period specified in the second indent of paragraph 1;
   (b) In the case of certain foodstuffs which keep for a long time, extend the period specified in the second indent of paragraph 1;
   (c) Without prejudice to the first indent of Article 23 (1) (b), in the case of foodstuffs which will keep for more than 12 months, extend to six years that period laid down in the second indent of paragraph 1 above as regards the obligation to indicate the date of minimum durability.
3. In the case referred to:
   (a) In paragraph 2 (a), the procedure laid down in Article 16 (2) shall apply to any national provision;
   (b) In paragraph 2 (b) and (c), Member States shall inform the Commission and the other Member States of any measure taken pursuant to the said points.

4. Member States shall also ensure that the Commission receives the text of any essential provision of national law which they adopt in the field governed by this Directive.

ARTICLE 23

...

ARTICLE 24

This Directive shall also apply to the French overseas departments.

ARTICLE 25

This Directive shall not apply to foodstuffs marketed in Greenland, intended for local consumption.

ARTICLE 26

This Directive is addressed to the Member States.

ANNEX I

Categories of ingredients which may be designated by the name of the category rather than the specific name

Definition
Refined oils other than olive oil
Designation
"Oil", together with
- -- either the adjective "vegetable" or "animal", as appropriate, or
- -- an indication of their specific vegetable or animal origin

The adjective "hydrogenated" must accompany the indication of a hydrogenated oil

Definition
Refined fats
Designation
"Fat", together with
- -- either the adjective "vegetable" or "animal", as appropriate, or
- -- an indication of their specific vegetable or animal origin

The adjective "hydrogenated" must accompany the indication of a hydrogenated fat

Definition
Mixtures of flour obtained from two or more cereal species
Designation
"Flour", followed by a list of the cereals from which it has been obtained, in descending order by weight

Definition
Starches, and starches modified by physical means or by enzymes
Designation
Starch

Definition
All species of fish where the fish constitutes an ingredient of another foodstuff and provided that the name and presentation of such foodstuff does not refer to a specific species of fish
Designation
Fish

Definition
All types of cheese where the cheese or mixture of cheeses constitutes an ingredient of another foodstuff and provided that the name of presentation of such foodstuff does not refer to a specific type of cheese
Designation
Cheese

Definition
All spices not exceeding 2% by weight of the foodstuff
Designation
Spice(s) or mixed spices

Definition
All herbs or parts of herbs not exceeding 2% by weight of the foodstuff
Designation
Herb(s) or mixed herbs

Definition
All types of gum preparations used in the manufacture of gum base for chewing gum
Designation
Gum base

Definition
All types of crumbed baked cereal products
Designation
Crumbs or rusks as appropriate

Definition
All types of sucrose
Designation
Sugar

Definition
Anhydrous dextrose or dextrose monohydrate
Designation
Dextrose

Definition
Glucose syrup and anhydrous glucose syrup
Designation
Glucose syrup

Definition
All types of milk protein (caseins, caseinates and whey proteins) and mixtures thereof
Designation
Milk proteins

Definition
Press, expeller or refined cocoa butter
Designation
Cocoa butter

Definition
All crystallized fruit not exceeding 10% of the weight of the foodstuff
Designation
Crystallized fruit

Definition
Mixtures of vegetables not exceeding 10% of the weight of the foodstuff
Designation
Vegetables

Definition
All types of wine as defined in Council Regulation (EEC) No 822/87
Designation
Wine

ANNEX II

Categories of ingredients which must be designated by the name of their category followed by their specific name or EC number

Colour	Flavour enhancer	Anti-foaming agent
Preservative	Acid	Glazing agent
Anti-oxidant	Acidity regulator	Emulsifying salts
Emulsifier	Anti-caking agent	Flour treatment agent
Thickener	Modified starch	Firming agent
Gelling agent	Sweetener	Humectant
Stabilizer	Raising agent	Bulking agent
		Propellent gas

ANNEX III

Designation of flavouring in the list of ingredients

1. Flavourings shall be designated either by the word "flavouring(s)" or by a more specific name or description of the flavouring.
2. The word "natural" or any other word having substantially the same meaning may be used only for flavourings in which the flavouring component contains exclusively flavouring substances as defined in Article 1 (2) (b) (i) and/or flavouring preparations as defined in Article 1 (2) (c) of Directive 88/388/EEC on flavourings.
3. If the name of the flavouring contains a reference to the vegetable or animal nature or origin of the incorporated substances, the word "natural" or any other word having substantially the same meaning may not be used unless the flavouring component has been isolated by appropriate physical processes, enzymatic or microbiological processes or traditional food-preparation processes solely or almost solely from the foodstuff or the flavouring source concerned.

As amended by
85/7/EEC: Council Directive of 19 December 1984
(Official Journal No L 2, 03/01/1985)
86/197/EEC: Council Directive of 26 May 1986
(Official Journal No L 144, 29/05/1986)

89/395/EEC: Council Directive of 14 June 1989
(Official Journal No L 186, 30/06/1989)
91/72/EEC: Commission Directive of 16 January 1991
(Official Journal No L 42, 15/02/1991)
93/102/EEC: Commission Directive of 16 November 1993
(Official Journal No L 291, 25/11/1993)
Council Decision 95/1 adjusting the instruments concerning the accession of new Member States (Austria, Finland and Sweden) to the European Union
(Official Journal No L 1, 01/01/1995)

APPENDIX F

1999 LEGISLATIVE PROPOSAL

Redefining "Perishable food" in Minnesota

With the passing of time comes advances in technology. While many technologies are potentially beneficial, they often times require changes or adjustments in traditional terminology. The definition used to define 'perishable food' in the State of Minnesota, for example, is now an outdated term due to advances in controlled atmospheric/modified atmospheric packaging techniques in the food industry.

Fresh fruits and vegetables, for example, are presently excluded from the 'perishable food' definition because of the variable nature of the fresh product and how it is handled in distribution. Controlled atmospheric/modified atmospheric packaging (CAP/MAP), however, is presently being used for many vegetable products. Consequently, this technology has outdated the current use of the word 'perishable food.' The following is the current definition, which excludes fresh fruit and vegetables, in Minnesota taken from Minnesota Statutes Annotated (1996) under 'Quality Assurance Dating':

31.782 Definitions.

> *Subd. 3. "Perishable food" means any food intended for human consumption (other than meat and poultry, frozen food, or fresh fruit or vegetables), which has a quality assurance date.*

The recent growth of the fresh pre-cut salad business (estimated to be greater than one billion dollars) makes adding 'a quality assurance date' on such products necessary. Pre-cut mixed products require the CAP/MAP technology to extend their shortened shelf-life due to senescence and microbial decay. This technology has also been applied to some meat products. Because of the short experience of grocers and consumers with these products, the 'quality assurance date' will help to move product from production to table more rapidly. This is urgent since many of the products may experience accelerated deterioration due to the fact that many grocery retailers keep their refrigerators at higher than recommended temperatures. Proper rotation will not prevent effects from temperature abuse, but it will assist is preventing longterm temperature abuse at the supermarket level. More rapid deterioration of the prepackaged foods at abused temperatures could lead to growth of pathogens, such as *Clostridium botulinum*, a potentially lethal microorganism. In one incident involving temperature abused CAP/MAP packaged shredded cabbage product, four people became ill with this microbial contamination.

To help promote the efficient rotation of the CAP/MAP products, these products must be considered a 'perishable food' in Minnesota legislation. The following modification in the definition of 'perishable food' is strongly recommended.

> *Subd. 3. "Perishable food" means any food intended for human consumption which has a quality assurance date. This definition excludes meat and poultry or fresh fruit or vegetables unless packaged under controlled atmospheric/modified atmospheric conditions. Frozen foods are also excluded.*

Your support and participation in correcting this matter are requested.

Lynn M. Szybist • Department of Food Science and Nutrition; University of Minnesota; 1354 Eckles Ave.; St. Paul, Mn 55108

APPENDIX G

Proposed Federal Open-Dating Regulation 21 CFR Part 101.19 Proposed Regulation

MANDATORY REGULATIONS REGARDING THE OPEN DATING OF FOOD PRODUCTS

By Theodore P. Labuza and Lynn M. Szybist

Summary:

An open date is the practice of labeling a packaged food with a date that indicates when the product was packed, meant to be "sold by", or meant to be "used by". The proposed regulation requires mandatory open dating of all food products. Such a regulation eliminates the need for individual state regulations, and it benefits consumers, retailers, and government agencies. Many consumers want open dates to help them make educated choices about the freshness of the foods they consume. A federally-mandated system will increase efficiency in stock rotation at the retail level and eliminate the present burden on interstate commerce caused by non-uniform state open-dating regulations.

Supplementary Information:

I. Background

A. Regulatory History

Open dating was established in the dairy industry in 1917 (Ref. 1). In the 1930s, Consumer Reports published consumers' desires for regulated open dating as a freshness indicator on their food products. As of the early 1970s, some supermarket chains independently began implementing some type of dating system (Ref. 2).

In 1979 the Office of Technology Assessment (OTA), the research branch of the US Congress committed to developing background papers, was charged with examining the effects and feasibility of mandatory open-dating legislation at the federal level (Ref. 3). The committee chairman was Massachusetts' Senator Ted Kennedy, whose state's statutes at that time required the open dating of all food products. A task force was formed which included consumer representatives, food retailers, processors, wholesalers, scientific experts, and state and federal government officials. OTA concluded that the industry lacked data to properly implement an efficient open-dating requirement at that time but recommended that the issue be addressed again.

B. Open Dating and Consumer Demands

Table 1 contains commonly used open-dating terminology and their definitions:

TABLE 1.
OPEN DATING TERMINOLOGY

Open-Dating Terminology	*Definition*
Production Date or Pack Date	Historical meaning; gives the date on which the product was manufactured or put into the final package. Used on prepackaged fresh fruits and vegetables, where shelf-life depends on the freshness of the product when harvested.
Sell-by Date	Helps in stock rotation to get the products out so the consumer can purchase the product at a point which will still give them adequate time for home storage before the end of shelf-life. Printed dates are usually very good guesses or industry practice based on assumed conditions.
Best-if-Used-by Date	The estimated point where the product quality loss reaches a level still generally acceptable but after which it fails to meet the high quality standard. Ambiguous date as to when the product should be taken off the supermarket shelf and confusing for the stock rotators.
Combination Date	"Best if used within __days of (date)." The "__ days of" part makes this phrase a "best-if-used-by" date, while the "date" given represents a "sell-by" date.
Use-by Date	Commonly interpreted as "it dies or you die if you eat it" (Ref. 4). The date determined by manufacturers as the end of useful quality life of the product.
Freeze-by Date	Often on meat or poultry in conjunction with another date, such as a use-by date. Helpful to the consumer and helps the store in terms of product movement.
Closed or Coded Date	Numbers used by the industry that indicate production lots. May represent a packing date, but not written for the consumer to understand. Important number for product identity in case of recalls.

(Ref. 3, 4, 5)

Studies have reported that consumers use the date as an indicator of shelf-life or safety and as a measurement of quality loss while retailers use the date as a means of product movement and stock rotation.

In 1973, an A.C. Nielsen Report found that many people look on the food package for some type of date to aid them in selecting fresh food (Ref. 6). A 1973 Economic Research Service (ERS) (USDA) and Consumer Research Institute (CRI) study on food spoilage reported a significant number of consumers were dissatisfied with the freshness of their foods, although 93% of those surveyed had not purchased any stale or spoiled food products within the past year. If the food was spoiled on the day of purchase, most consumers discarded the product despite the fact that 62% of those shoppers knew about the store money-back guarantee (Ref. 4). Also in the 1970s, the New York State Consumer Protection Board (a consumer group) published a book deciphering the meanings of manufacturers' code dates, which were not consumer-friendly and were used as a means of lot identification in case of product recalls. Over 100,000 orders were placed for this code book in its first year alone indicating a high interest in the dating of food products (Ref. 7). Finally, in 1979, the Food and Drug Administration (FDA) conducted a survey of 1,374 grocery shoppers. FDA reported that 94% of those surveyed claimed to have noticed the open dating on some food products, and 75% of them have used the date in making a purchase (Ref. 7).

The popularity of the open-dating issue seemed to lessen after 1980 until about 1996. At that latter time, *Supermarket News*, a publication targeted at the grocery industry, reported that based on a consumer study, peak freshness was the most important quality consumers looked for and that seeing a "sell-by date" on a food product heightened consumer confidence about that product's freshness (Ref. 8). Joanne Gage, Vice President of Consumer and Marketing Services in Schenectady, NY, reported that increasing effective communications with the consumer required simple and basic information. The "Sell-by" or "Use-by" dates on packaged items, such as fresh-cut salads, were highly sought after by the consumer (Ref. 9).

C. Current Policies

1. Federal Regulations

The only federally controlled open-dating policies are for prescription and over-the-counter drugs and infant formulas (21 CFR 107.20 (c)). Meat, poultry and whole uncracked eggs are regulated by the United States Department of Agriculture (USDA). Although USDA also does not mandate an open- dating system on meat, if a date is printed on a package it must be accompanied with an explanatory phrase (Ref. 7). More recently, FDA, in an advanced notice of proposed rulemaking, proposed some form of mandatory dating for dietary supplements (Ref. 10) as a good manufacturing practice (GMP). One should note that Section 201 (n) of the Federal Food, Drug and Cosmetic Act denotes that a label can be deemed misbranded if "it fails to reveal facts material in the light of such representations." Based on this, one can make the case that since many foods deteriorate over time, an open date is a critical material fact that the consumer should have on the label.

2. National Institute of Standards and Technology (NIST)

NIST is the federal agency responsible for creating uniformity among states concerning local weights and measures laws, standards and practices. The agency has also attempted to standardize the open-dating policies among the states as published in the Uniform Regulation NIST Handbook #130. The Uniform Regulation provides states with two open-dating options: (1) require open dating on all perishable foods, or (2) allow the voluntary usage of open dating of perishable foods as regulated in the handbook (Ref. 11). By 1997, 16 out of 50 states and several territories had adopted some form of the NIST model regulation (Ref. 11).

3. State Regulations

In 1979, OTA reported that 22 states mandated some sort of open- dating policy. Unlike NIST, which communicated only with weights and measures offices within each state, OTA contacted the appropriate department in each state that would handle open dating. In 1998, 30 states and the District of Columbia regulated an open-dating policy (Ref. 12) as shown in Table 2.

TABLE 2.
PAST AND CURRENT OPEN-DATING REGULATIONS BY STATE

	As of 1979			As of 1998	
State	*Primary Products*	*Form of Open Date*	*Effec-tive Since About*	*Primary Products and Required Form of Open Date*	*Source*
Alabama	Dairy	Sell-by	1975	Potentially Hazardous Foods	Department of Agriculture and Industries Agricultural Chemistry: ss. 80-1-22-.28
Alaska				Voluntary	
Arizona				Expiration Date on Eggs	ALIS Online: ss. 3-719(E) (http://www.azleg.state.az.us/cgi-bin/wais)
Arkansas				Voluntary	
California	Dairy	Sell-by	1973	Milk and Milk Products	California Annotated Codes (1986): Division 15, ss.36004
Colorado				Voluntary	

TABLE 2. (Continued)

	As of 1979			As of 1998	
State	***Primary Products***	***Form of Open Date***	***Effective Since About***	***Primary Products and Required Form of Open Date***	***Source***
Connecticut	Milk	Sell-by	1973	Milk and Milk Products	Connecticut General Statutes Annotated (1985): ss. 22-197b
Delaware				Voluntary	
District of Columbia	Perishable Products	Sell-by	1974	Perishables	District of Columbia Municipal Regulations (1997): Title 23, ss.2505
Florida	Dairy	Sell-by	1976	Milk and Milk Products	Florida Statutes Annotated: ss 502.042
Georgia	Milk and Eggs	Sell-by	1973	Food Products in Package Form	Rules of Georgia Department of Agriculture: ss. 40-7-1-.26
Hawaii				Milk and Milk Products	Hawaii Administration Rule: ss. 11-15-39 (c)
Idaho				Voluntary	
Illinois				Voluntary	
Indiana				Voluntary	
Iowa				Refrigerated Foods in Reduced Oxygen Packages	Iowa Code (1997): 481-31.11 (137A)
Kansas				Voluntary	
Kentucky				Milk and Milk Products	Department for Health Services - Open Dating Requirements for Milk Products (902 KAR 50:080)
Louisiana				Voluntary	
Maine				Voluntary	
Maryland	Milk	Sell-by	1971	Milk Products	Annotated Code of MD (1996) (1997 Supplement): ss. 21-426
Massachusetts	Perishable & Long Shelf Life	Sell-by or Use-by	1979	Perishable and Semi-Perishable Food Products	Code of Massachusetts Regulations 105 CMR: 520.119

TABLE 2. (Continued)

	As of 1979			As of 1998	
State	***Primary Products***	***Form of Open Date***	***Effective Since About***	***Primary Products and Required Form of Open Date***	***Source***
Michigan	Perishable Products	Sell-by	1969	Milk and Milk Products	MI Dept of Agriculture, Dairy Division, Regulation No. 408. (http:/ / www.state.mi.us/ execoff/admincode)
Minnesota	Perishable Products w/Shelf Life < 90 Days	Sell-by or Use-by	1973	Perishable Products with Shelf Life < 90 Days	Minnesota Statutes Annotated (1996): ss. 31.781 - 31.784
Mississippi				Eggs	Mississippi Egg Marketing Law and Regulations-Regulation 1
Missouri				Voluntary	
Montana				Fluid Milk and Grade A Milk Products	Administrative Rules of Montana: ss. 32.8.201 - 32.8.205
Nebraska	Eggs	Pack	---	Eggs & Refrigerated Reduced Oxygen Packaged Food	Food Establishment: Reduced Oxygen Packaging Method: ss. 81-2,272.27. (http:// www.agr.state.ne.us) & Nebraska Graded Egg Act and Rules & Regulations
Nevada	Dairy	Sell-by	1973	Voluntary	
New Hampshire	Cream	Use-by	1973	Refrigerated Prewrapped Sandwich & Prepackaged Perishable Foods	NH Department of Agriculture, Markets & Food: Agriculture ss. 1413.04 and NIST Handbook
New Jersey	Dairy	Sell-by	---	Fluid Milk	New Jersey State Department of Health (1994): ss.8:21-10.1
New Mexico	Milk	Sell-by	1977	Milk and Milk Products	http:// www.nmdaweb.nmsu. edu
New York				NYC - Milk and Milk Products	NYC Health Code: ss. 111.33

TABLE 2. (Continued)

	1979			1998	
State	***Primary Products***	***Form of Open Date***	***Effective Since About***	***Primary Products***	***Source***
North Carolina				Smoked Fish	NC Administrative Code: ss. .0507e(5)
North Dakota				Voluntary	
Ohio	Perishable Products	Sell-by	1977	Perishable Products	OH Department of Agriculture (Regulation 901:3-57-04)
Oklahoma	Meat, Eggs	Sell-by	---	Voluntary	
Oregon	Perishable Products	Pack or Sell-by	1975	Packaged Perishable Products	Title 49: Chapter 616, ss. 616.815 - 616.830
Pennsylvania	Milk	Sell-by	1975	Milk	PA Code: Title 7, Part III: ss. 59.22
Rhode Island				Voluntary	
South Carolina				Voluntary	
South Dakota				Expiration Date on Eggs	South Dakota Dept. of Agriculture: Law 39-11: ss.12:26:10
Tennessee				Voluntary	
Texas				Voluntary	
Utah				Voluntary	
Vermont				Voluntary	
Virginia	Dairy and Infant Formula	Sell-by	1974	Milk and Milk Products	ss. 2VAC5-490-40 (http://leg1. state.va.us/000/ reg.TOC02005. HTM#C0490)
Washington	Dairy and Others	Sell-by	1974	Perishable Packaged Foods	Commerce in Food, Drugs, and Cosmetics RCW 69..04.900-69.04.905
West Virginia				Prepackaged Perishable Foods	West Virginia Code (1996):ss. 47-1-9 and NIST Handbook
Wisconsin	Smoked Fish	Pack	1971	Smoked Fish	Wisconsin Administrative Code (1998): ATCP 70.22
Wyoming				Voluntary	

(Ref. 3, 12)

Table 3 represents the differences among the states pertaining only to present pasteurized milk dating regulations. These contradictions demonstrate a definite burden on interstate commerce.

II. Benefits of Open Dating

A. Food Safety

A regulated open-dating system serves as a guide for "first in-first out" practices at the retail level. In 1999, the Centers for Disease Control (CDC) reported 21 deaths (15 adults and 6 miscarriages/stillbirths) across the nation from an outbreak of *Listeria monocytogenes* (Ref. 13). The source of the foodborne illness was ready-to-eat meat products, such as hot dogs and luncheon meats, consumed very close to the stamped "use-by" dates. If the products were temperature abused, however, conditions may have allowed the pathogenic organisms to achieve an infectious level before the end of other food quality attributes (Ref. 14). To avoid such incidents in the future, federal rules governing how the "sell-by" and "pull-by" dates are determined on packaged foods is necessary (Ref. 15). A shorter shelf-life on such products will increase the products' turn-over rates and lessen the chances of microbial infections.

TABLE 3.
OPEN DATING ON PASTEURIZED MILK PRODUCTS

State	Sell-by/Pull Date
California	Established by the processors/manufacturers
Connecticut	12 day maximum
Florida	Established by the processors/manufacturers
Hawaii	Established by the processors/manufacturers
Kentucky	Established by the processors/manufacturers
Maryland	14 day maximum
Michigan	Established by the processors/manufacturers
Montana	12 day maximum
New Jersey	Established by the processors/manufacturers
New Mexico	Established by the processors/manufacturers
New York City	96 hour maximum (pasteurized milk); 15 day maximum (ultra-pasteurized milk)
Pennsylvania	14 day maximum
Virginia	96 hour minimum; maximum number of days established by the processors/manufacturers

(Ref. 12)

B. Overall Quality

Under proper conditions, an open date can indicate the end of acceptable food quality and increase consumer confidence in the food products they purchase. Out-dated foods are more likely to have lower sensory qualities relating to color, texture, odor and flavor changes (Ref. 3). In addition, as stated in Grocery Manufacturers of America (GMA) v. Department of Public Health (393 NE 2d, 881, 1979), the dating of foods helps deter the possible consumption of old products which may have a higher risk for non-safe agents (microorganisms).

C. Nutrient Levels

Based on the temperature and humidity conditions during distribution, degradation of several essential amino acids and vitamins (such as A, B, and C) may cause some food products to have nutrient levels below the level stated on the label (Ref. 7). It is not clear how companies base their open dates in compliance with nutritional labeling laws as defined in 21 CFR 101.9 (g).

D. Functional Properties

During storage, some foods will lose their functional properties (Ref. 16). Baking yeast, for example, will lose some of its leavening capabilities; eggs will decrease in their whipping abilities; and pre-emulsified foods, such as salad dressing, may separate over time, especially if temperature abused. With respect to eggs, it is suggested that they will fall below the labeled grade (A, B, etc.) within 30 days at refrigerated temperature.

E. Retail Stock Control

In 1972, Keith Ford of the Minnesota Office of Consumer Services found that in a survey across the State of Minnesota, 44% of the infant formulas on grocers' shelves were past the use-by date. With more than a hundred thousand different code dates being used by the food industry, another study found that 64% of store managers could not even read the date codes on food packages (Ref. 4). The current dating practices continue to be confusing to retailers and consumers as shown in Table 4 with respect to yogurt (Ref. 12). The table also demonstrates various dating systems among yogurt manufacturers and differences within the same brand in regard to the suggested storage temperature. There was no explanation of the date on container of Dannon yogurt. To understand the date, the individual must dial the toll-free number on the container, which is inconvenient for the consumer and the stock rotator at the retail level.

F. Home Stock Control

A survey was conducted in 30 households to evaluate the home storage practices of participants (Ref. 17). The problems of careless storage practices can lead to increased customer dissatisfaction with the product, stale and rancid products, and the presence of insects. It might also result in a food safety problem. Table 5 shows some of the survey results on home storage practices.

The results indicate that the home storage rotation practices of the participants were poor, especially with nonperishable items, such as baking products, dry goods, condiments, and breakfast cereals. Since many households did not date their purchases, an open date would be useful in managing household stocks (Ref. 17).

TABLE 4.
CURRENT OPEN-DATING PRACTICES ON YOGURT CONTAINERS
(Collected on 1/13/98)

YOGURT		
Product & Address	*Open Date*	*Printed Dates*
(5 oz) **Kemps Nonfat** (Various Flavors) **Yogurt** Marigold Foods, Inc./General Offices Mpls., MN 55414	Remains wholesome one week after date on carton.	JAN.28 to FEB.17
(32 oz) **Kemps Nonfat Vanilla Yogurt** Marigold Foods, Inc./General Offices Mpls., MN 55414	When properly refrigerated between 33 & 44°F, this product will retain its wholesomeness for 1 week beyond date on carton.	FEB.05
(8 oz) **Gaymont Lowfat** (Various Flavors) **Yogurt** (Old Home Foods) Old Home Foods St. Paul, MN 55103	Quality assured 7 days beyond code date on side if properly refrigerated (40-44).	FEB.05 to FEB. 19
(24 oz) **Gaymont Lowfat** (Various Flavors) **Yogurt** Old Home Foods St. Paul, MN 55103	Quality assured beyond code date on side if properly refrigerated (35-40).	FEB.5 to FEB.12
(8 oz) **Dannon Yogurt** Dannon Company, Inc. Tarrytown, NY 10591	---	JAN.17 to FEB.8

(Ref. 12)

TABLE 5.
STORAGE TIME OF PRODUCTS IN CONSUMER PANTRIES

PRODUCTS	STORAGE TIME (WEEKS)	
	Average	Range
Canned Goods	12.3	1-104
Ethnic Foods	10.9	1-52
Condiments	11.6	1-156
Baking Products	21.6	1-260
Dry Goods/ Pasta, Mixes	17.7	1-156
Breakfast Cereals	12.0	1-150
Cookies, Crackers, Snacks	7.6	0.3-26
Coffee, Juices, Beverages	36.3	1-156
Dog/ Cat Food	11.0	1-28
Paper Products	--	1-52

(Ref. 17)

G. Educating the Consumer

A 1979 FDA study found that only 1.3% of respondents felt confused about what the open date on food packages represented. After further questioning, however, it was revealed that actually most of those surveyed did not understand the meaning of the date (Ref. 7). A 1979 OTA study gathered the following information concerning consumers' understanding of open dates.

TABLE 6.
CONSUMER UNDERSTANDING OF FRESHNESS DATES
(Percentage of Respondents)

	MILK	*BREAKFAST CEREAL*	*GROUND BEEF*
When it was packaged...	9	8	***34****
Last day it should be sold...	***74****	35	31
Last day it should be used or eaten...	15	***26****	9
Have never noticed a date on a package of this product...	2	31	26

Note: Percentages in boldface with asterisks indicate correct answers (Ref. 3).

The participants were mixed as to what they understood the date to mean on milk, breakfast cereal, and ground beef packages. About three-quarters of shoppers correctly identified the date on milk; only one-quarter of the answers for breakfast cereal and one-third of the answers for ground beef were correct. A 1992 Minnesota/South Dakota Dairy Center study showed that 94% of those surveyed thought the open date was extremely important. But 25% of that population doubted the reliability of the date, while 61% admitted that they did not fully understand it (Ref. 18).

H. Decrease Overall Misconceptions

In an Institute of Food Technologists (IFT) press release to newspaper editors (December 15, 1998), some "food safety" tips were given, such as the following statement: "**Don't debate the date.** Don't buy food past its expiration or sell-by date, or food that will not be used by its sell-by date" (Ref. 19). The tip implies that all products consumed before their dates are safe, which is untrue if the food is temperature-abused in transportation or storage. It also implies that the food becomes unsafe at the end of the sell-by or use-by date. Manufacturers generally base their dates on quality not safety, especially with respect to the sell-by date, which was intended to help retailers in stock rotation. Therefore, this recommendation, while useful, is not totally truthful. Only by knowledge of the time-temperature history during distribution, would one be able to make some judgement on safety.

Recently, a study by FDA (Ref. 20) for potential outgrowth of *Clostridium botulinum* and toxin production in vacuum packaged fish, suggested that time-temperature integrating tags (TTIs) would be very useful for monitoring the storage abuse on individual packages. In fact, the 1979 OTA study (Ref. 3) suggested that research be done on TTIs and in 1989 Taoukis *et al.* published an overview of the then current status of TTIs (Ref. 21). In 1998, CUB Foods in the upper Mid-West introduced TTIs on its prepackaged raw hamburger in a test market of the concept.

I. Reducing 'Out-Dated' Legislation

The open-dating legislation is outdated in several states. In Minnesota, for example, perishable foods, excluding fresh fruits and vegetables or meat and poultry, are required to have an open date. However, with advances in technology, certain foods, including fresh fruit and vegetables and meat can be packaged under controlled atmospheric/modified atmospheric (CAP/MAP) conditions. This process increases the shelf-life of the foods, but also adds a possible presence of anaerobic growths if the food package is mishandled. Because of the short experience with these foods among retailers and consumers, ensuring proper rotation is necessary to help move the product from production to table more rapidly. An open date must be mandated for CAP/MAP foods regardless of the food content. Other out-dated legislation includes the open-dating laws of Iowa and Nebraska, which require the dates on reduced oxygen packaged foods (such as CAP/MAP packaged foods) not to exceed 14 days. Adhering to these laws would make some refrigerated filled-pasta products illegal in those states if held for sale beyond this time (even though they have about a three month shelf-life under normal refrigerated conditions).

III. Public Health Importance

Grocery Manufacturers of America (GMA) v. Department of Public Health (393 NE 2d, 881, 1979) was the only open-dating case to ever reach a court decision. Massachusetts passed regulations to require food vendors to disclose a "last date of use" or "pull date" on all food products. GMA proclaimed that Massachusetts' legislation created an unnecessary burden on commerce. The State's Supreme Court, however, decided in favor of the Public Health Department ruling that the regulation "bore reasonable relation to goal (sic) of consumer protection" because out-of-date products have an increased risk for non-safe agents. The Department had the authority to regulate the sale of food with respect to labels that could be "misleading in any particular" including "labels containing omissions of fact as well as sale of food with labels containing express misstatements of fact." The court stated that a food label which is "misleading in any particular" makes the food product misbranded. Therefore, a food product with no date was misbranded since that would be an omission of fact and the purpose of the date was to protect public health.

IV. Economic Impact

OTA accessed the average cost of food shelf-life testing in their 1979-80 report. The price to determine the shelf-life of a perishable food was calculated to be approximately $100,000 per food item. For shelf-stable products, the cost was about $200,000. Shelf-life testing would be considerably higher today. The total costs would also increase with additional testing necessary with changes in product formulation, changes in packaging and with inflation considerations. There would also be a financial consideration for the printing technology needed to apply a date. Enforcement costs would depend on the enforcement system proposed to be used (Ref. 3).

V. Environmental Impact

Open dating may restrict the ability to sell products that remain 'fresh' but have reached their 'sell-by' date. According to USDA, 5.4 billion pounds of food at the retail level were estimated to be wasted in 1995, and 91 billion pounds were lost at the food service and home level (Ref. 22). Fluid milk waste was estimated to be almost 30% (Ref. 22). Dairy and bakery products are two of the largest contributors to food loss. Such products, which may still be safe to eat and of acceptable quality, are often removed from supermarket shelves once reaching the 'sell-by' date because many people relate the open date as the last date to safely sell the product or the last date to safely eat it (Ref. 23). In a 1971 study, 62% of 628 people in the survey stated that they sometimes sort through packages to find the freshest product. From that same group, 74% claimed that while sorting through dated products, they would usually find some products that were fresher than others. This out-of-date food in part goes to food banks, which is a benefit, but a lot of food waste ends up in landfills. Mandatory open dating on a national scale could increase waste, however, if it impacts by forcing better distribution control and grocery store shelf stock rotation, it might in fact have a positive impact on decreasing waste.

VI. Risk Analysis

A. Temperature Abuse

As stated by Dr. Ted Labuza, "Shelf-life is not a function of time alone, rather it is a function of the environmental conditions and the amount of quality change that can be allowed" (Ref. 4). These "environmental conditions" relate to the temperature of food products during storage and distribution.

The maximum temperature recommended for chilled foods in warehouses, trucks, and retail displays by the National Food Processors Association (NFPA) is 40°F (Ref. 24). NFPA has also published the following information for the optimum temperature ranges of various food products during distribution and storage:

TABLE 7.
OPTIMUM TEMPERATURES FOR CHILLED FOODS

Food Product	***Optimum Temperature Range***
Dairy Products	32-40°F
Meat	30-34°F
Poultry	30-34°F
Seafood	30-34°F
Salads	32-40°F

(Ref. 24)

In the revised Food Code (1997), FDA established 41°F as the recommended maximum temperature for retail establishments which handle meat, fish, poultry, delicatessen products and precut produce (Ref. 24).

Besides the possible expense of having to employ new technology to maintain proper temperatures, retailers and distributors face the financial responsibility of a 10% increase in total energy costs for every 5°F their refrigeration level is lowered (Ref. 25). The economic costs should not overstep the boundaries of safety, but studies show that the safety issue is not being taken seriously and proper steps in temperature control are not being enforced. The following examples demonstrate the problem of temperature abuse.

1. 1989 Audits International Survey

Temperature measurements were recorded of 1,000 refrigerated food items at three points: at the retail level, when the products reached the consumers' homes, and after 24 hours. Delicatessens in the study had refrigerated foods stored at temperatures ranging from 34 to 71°F; the average was 47°F. The average home refrigeration was 43°F (Ref. 26). Both of these averages are above the recommended 40-41°F temperatures for refrigerated food, meat, poultry, and eggs.

2. 1998 Audits International Study

Almost 10 years later, Dr. Daniels of Audits International reported a follow-up study at the International Fresh Cut Produce Conference in San Diego (Ref. 27). In a sample of 98 supermarkets, the mean temperature in refrigerated deli cases was 46°F with 10% of them at a mean of 58°F. In the produce section, packaged salads were also at a mean of 46°F. In both of these cases unacceptable temperatures will cause more rapid deterioration of prepackaged foods and could lead to a safety issue.

3. 1997 The Refrigeration Research and Education Foundation

The study reported that approximately 20% of retail chilled display cabinets operate at temperatures above 50°F, and that over two-thirds of the chilled food retailers also do not monitor the expiration dates on their products (Ref. 24). These studies illustrate a potential for growth of pathogens that might cause food poisoning especially in abused chilled food products that are minimally processed. At temperatures above 40°F, pathogenic anaerobic microorganisms are capable of growth and toxin production. Between the temperatures of 40 and 55°F, nonproteolytic anaerobic microorganisms may grow. Such spoilage is not detectable by virtue of smell (Ref. 28).

4. 1998 The University of Nottingham, England

Led by Dr. Angela Johnson, this study concentrated on the elderly, who represent the highest risk of mortality from food poisoning (Ref. 29). The participants' food safety knowledge and practices were evaluated. While the results for overall understanding of the "sell by" and "use by" dates were decent, 45% of them could not read the dates because of poor eyesight and because of small and hard-to-read print. Of even greater concern, however, was the fact that 70% of participants' refrigerators were too warm to safely store food.

5. 1992 Vaccinations and Temperature Abuse

Lack of temperature control is not just a problem in the food distribution chain. Bishai *et al.* (1992) found that of 50 pediatric clinics in the Los Angeles area, only 16% of vaccine (measles, mumps, rubella) storage coordinators could cite appropriate storage temperatures for vaccines and 18% were unaware that temperature abuse would destroy the effectiveness of the vaccines (Ref. 30). Refrigerator thermometers were checked once weekly in only 20% of offices, 22% of refrigerators were above the required temperature range, and 16% of offices stored vaccines unrefrigerated.

B. Transportation and Other Concerns

In 1995, because of these concerns with regard to lack of proper refrigeration, a Transportation Analysis Group (TAG) was formed by the U.S. Department of Transportation (DOT) and the Food Safety and Inspection Service (FSIS) of USDA to discuss potential solutions based on the HACCP concept. After an open meeting in Washington DC in June of 1996, FDA announced an advanced notice of proposed rulemaking jointly with FSIS and DOT (Ref. 31). In this document, they note that "...post processing transporters, storage operators and retail stores, restaurants and other food service sectors are important links in the chain of responsibility for food safety." Because of this and the noted lack of temperature control, these regulatory agencies proposed six possible steps that could be taken. Note that this includes holding at the grocery store level:

1) Setting mandatory temperature performance standards, e.g., a maximum of 41°F or 45°F for potentially hazardous foods.
2) Requirement of shipper record keeping.
3) A mandatory HACCP-type system.
4) Voluntary guidelines such as in the AFDO "Guidelines for Transportation of Food."

5) A combination of approaches.
6) No federal initiative.

All of these alternatives would have an impact on open dating but as of present, there has been no proposed regulation. Note that in 1990, a comment by Dr. Ted Labuza of the University of Minnesota in regard to a USDA/FSIS advanced proposed rulemaking on refrigerated foods suggested the use of time-temperature integrator (TTI) tags (Ref. 32). TTIs are one method that could potentially solve part of the problem of temperature abuse for chilled foods and would fall under the HACCP approach above (Ref. 32).

An open date will not protect consumers from microbial threats, but it can be useful as a guide. If the product is kept at ideal temperatures and conditions throughout its life-span, then the food will most likely be safe. In 1998 an article in *Newsweek* reported that to guarantee that temperature abuse has not occurred, "tell the truth" tape or time-temperature integrators (TTIs) were being designed to signal a premature end of shelf-life (Ref. 21, 33). After consumers check the TTI, they have confidence that their product was not temperature-abused, and the open date indicates approximately how long the product will remain fresh under proper conditions. Such information is useful at both the retail and consumer levels, since many consumers are not aware of the actual storage conditions in their own homes and local supermarkets.

1. 1991 Nabisco Food Company

A task force found that most consumers are not aware of the importance of keeping a proper refrigerator temperature and its effect on food stability and ingredient performance (Ref. 17). They discovered that out of 14 home refrigerators and 11 freezers they checked, only 7 refrigerators and 1 freezer had a thermometer. In the refrigerators, temperatures ranged from 32°F to 55°F; in freezers, the temperatures ranged from 5°F to 20°F.

2. 1980s-1991 Tropicana Products, Inc.

In the 1980s, the company discovered that 72% of consumer complaints on chilled juices were connected with temperature abuse (Ref. 34). When examining this problem, it was found that the temperatures during distribution of the juice reached 45°F in some cases. The recommended temperature range for proper storage of juice is 32-38°F to maintain high quality. Within the retail stores, only 37% of products were stored at the proper temperatures. During storage, the average temperature of juice was about 44°F and went as high as 56°F. Rotation practices in the display cabinets were also found to be poor.

In 1991, the company conducted a time-temperature experiment measuring the flavor quality. The juice was rated on a scale of 1-9 with scores below 5 being unacceptable. The results showed dramatic differences in quality with temperatures only 10°F different. These changes of shelf-life as a function of time-temperature are a function of its temperature sensitivity factor (Q_{10}). Juice held at 45°F fell to a 5.1 score after only 49 days, while juice held at 35°F, i.e., 10°F lower, remained above 6.0 for the entire 63-day shelf-life (Ref. 34).

VII. Time-Temperature Indicators

A. Definition and Potential for Use

Loss of quality and "freshness" are not solely a function of time, but depend on the control of temperature, humidity and light during distribution. Both the kinetics of deterioration and the product's history of exposure, including temperature abuse, must be incorporated into shelf-life dating (Ref. 3). Due to such obstacles, the OTA Committee did not pursue a federally mandated open-dating legislation.

With the proper implementation of an open date along with a time-temperature integrating tag (TTI), the degradation of a food product is monitored as a function of both time and temperature (Ref. 33). TTIs can increase the effectiveness of quality control in distribution and during storage (Ref. 21), because of its ability to indicate a shorten shelf-life due to temperature abuse. The open date gives the consumer a sense of the shelf-life left for a product. But if the TTI indicates the end-of-shelf-life much earlier than the open date, there may be a distribution or storage problem of which the manufacturer, retailer and consumer may not have been aware.

TTIs are small, physical devices that are placed on the food package to measure the temperature history of a product and indicate a definitive change at the end of shelf-life. Through "integration" of the time-temperature exposure, TTIs are reliable indicators of end of shelf-life for food products if they have similar temperature sensitivities (Q_{10}) as for the deterioration mechanism (Ref. 21). The devices can be used on individual consumer packages, so they establish a control system because not all products will receive uniform handling, distribution and time-temperature effects (Ref. 21). Therefore, TTIs can increase the effectiveness of quality control in distribution, stock rotation practices of perishable foods in grocery stores, and efficiency in measuring freshness by the consumer (Ref. 35). Commercially available TTIs are both reliable and applicable for use in combination with open dating of refrigerated foods (Ref. 36,37). One study showed that although the tags respond more quickly to temperature abuse than the actual food, the response is on the conservative side of safety, i.e., the tag color indicates end of shelf-life before the food is actually spoiled (Ref. 38). The Campden Food and Drink Association in the United Kingdom has developed technical standards for evaluation of TTIs (Ref. 39).

In 1988, Find SVP, a marketing consulting firm, concluded that the use of TTIs had great potential for use in the food business (Ref. 40). A survey by the Business Marketing Research Inc. found that consumers would prefer clear, consumer-readable indicators of time and temperature to measure the freshness and safety of their perishable food products (Ref. 18). In 1982, one study indicated that the 3M and i-point products were found to be very effective and "makes it possible to replace the sell-by date on market milk" (Ref. 41). A second study concerning the effectiveness of TTIs and the dairy industry found the tags to be beneficial in conjunction with an open date (Ref. 42).

B. Food Safety

TTIs are recommended to monitor the potential danger of temperature abuse for controlled atmosphere/modified atmosphere packaged (CAP/MAP) prepared meals. Improper conditions can lead to the growth of harmful pathogens such as *Clostridium botulinum, Listeria monocytogenes*, and *Salmonella enteritidis*. In one reported case, four people became ill from *C. botulinum* (a potentially lethal pathogen) from a CAP/MAP packaged shredded cabbage product that experienced temperature abuse (Ref. 17). FDA has shown that TTIs would also be very beneficial for CAP/MAP fish (Ref. 20).

VIII. Stake Holders

Everyone who buys market food in this country would be affected by a federal open-dating policy, however, the food industry and the government are the biggest stake holders. The passing of a mandatory open-dating system requires the financial responsibility of conducting shelf-life tests for the food industry. This increases consumer confidence in their food products and decreases the chances of consumer dissatisfaction from old products with unacceptable sensory qualities or with food safety issues. The government's responsibility involves monitoring or setting standards for the industry to follow when determining the open date of their product(s) and providing financial support towards widespread education of open dating through the media.

Retailers must ensure that the food is being properly rotated and that out-of-date products are removed from the shelves. Finally, the consumer's personal responsibility involves monitoring the open dates on individual purchases and becoming informed as to the proper use and meanings of open dates.

IX. References:

(1) Minneapolis Tribune, "Open Dating and Food Waste," *Cereal Foods World*, pp. 504–513, October 1979.

(2) Seligsohn, M., "Smashing the Open-Dating Myth," *Food Engineering*, pp. 20-25, October 1979.
(3) Office of Technology (OTA), Open Shelf-Life Dating of Food, Washington DC: U.S. Government Printing Office, 1979.
(4) Labuza, T.P., Shelf-life Dating of Foods, Trumbull, Ct: Food & Nutrition Press, Inc., 1982.
(5) Food Safety and Inspection Service (FSIS), URL: www.fsis.usda.gov, March 1995.
(6) A.C. Nielsen Co., "Study of Consumer Attitudes Toward Product Quality," Northbrook, Il, 1973.
(7) Institute of Food Technologists (IFT), "Open Shelf-life Dating of Food," *Food Technology*, pp. 89–96, February 1981.
(8) Dowdell, S., "Looking for a Date," *Supermarket News*, pp. 27–36, June 17, 1996.
(9) Williams, M., "Simple Advice," *Supermarket News*, pp. 31–2, July 13, 1998.
(10) Federal Register, "GMPs for Dietary Supplements," Vol. 62, pp. 5700–5709, February 8, 1997.
(11) National Institute of Standards and Technology (NIST), (http:www.nist.gov/) and Handbook 130 of the Federal Register.
(12) Labuza, T.P. and L.M. Szybist, "Current Practices and Regulations Regarding Open Dating of Food Products," Working Paper #99–01 Sloan Foundation, The Retail Food Industry Center, March 1999.
(13) Centers for Disease Control (CDC), "Update: Multistate Outbreak of Listeriosis," URL: http://www.cdc.gov/od/oc/media/pressrel/r990114.htm, March 1999.
(14) Silverman, J., Interview Food Safety Report ISSN 1523–4533, Bureau of National Affairs, Washington, D.C., February 17, 1999.
(15) Anon., "Bad Politics, Bad Meat," *The Los Angeles Times*, February 12, 1999.
(16) Rutgers University Food Science Department, Food Stability and Open Dating, Rutgers University: New Brunswick, NJ, 1971.
(17) Beard III, T.D., "HACCP and the Home: The Need for Consumer Education," *Food Technology*, pp. 123, June 1991.
(18) Sherlock, M. and T.P. Labuza, "Consumer Perceptions of Consumer Type Time-Temperature Indicators for Use on Refrigerated Dairy Foods," *Dairy, Food and Environmental Sanitation*, pp. 559–565, August 1992.
(19) Institute of Food Technologists (IFT), "Open Shelf-life Dating of Food," *Food Technology*, pp. 89–96, February 1981.

(20) Skinner, G.E. and J.W. Larkin, "Conservative Prediction of Time to *Clostridium botulinum* Toxin Formation for Use with Time-Temperature Indicators to Ensure the Safety of Foods," *J. Food Protection* 61: 1154–1160, 1998.
(21) Taoukis, P. and T.P. Labuza, "Time-Temperature Indicators," *Food Technology*, pp. 70–82, October 1991.
(22) Kantor, L.S., K. Lipton, A. Manchester, and V. Oliveira, "Estimating and Addressing America's Food Losses" *FoodReview*, pp. 2–12, January-April 1997.
(23) Anon., "Prevention Magazine-NBC Today Survey," 1997.
(24) Brody, A.L., "Chilled Foods Distribution Needs Improvement," *Food Technology*, pp. 120, October 1997.
(25) FNQUEB Webpage, URL: http://www.fnqeb.com.au/index.html.
(26) Richard Daniels Associates, Inc., "1989 National Retail Food Product Cold Temperature Evaluation," Audits International (USA), 1990.
(27) Anon., "Retail Temperature Control Lags Food Code": IFPA Seminar Speakers in *Food Regulation Weekly*, Vol. (2):3–4, 1998.
(28) Brody, A.L., "Minimally Processed Foods Demand Maximum Research and Education," *Food Technology*, pp. 66, May 1998.
(29) Johnson, A.E., A.J.M. Donkin, K. Morgan, J.M. Lilley, R.J. Neale, R.M. Page, and R. Silburn, "Food Safety Knowledge and Practice Among Elderly People Living at Home," *Journal of Epidemiology and Community Health*, pp. 745–748, November 1998.
(30) Bishai, D.M., S. Bhatt, L.T. Miller, and G. Hayden, "Vaccine Storage Practices in Offices of Pediatric," *Pediatrics* 88:193–196, 1992.
(31) Federal Register, "Transportation and Storage Requirements for Potentially Hazardous Food," Vol. 61, pp. 59372–59382, November 22, 1996.
(32) Anon., "Time temperature devices USDA requirement urged," *Food Chemical News*, pp.36, October 15, 1990.
(33) Springen, K., "Safer Food for a Tastier Millennium," *Newsweek*, pp. 14, September 28, 1998.
(34) Kalish, F., "Extending the HACCP Concept to Product Distribution," *Food Technology*, pp. 119, June 1991.
(35) Sherlock, M., B. Fu, P.S. Taoukis, and T.P. Labuza, "A Systematic Evaluation of Time-Temperature Indicators for Use as Consumer Tags," *Journal Food Protection*, 54:885–889, November 1991.
(36) Taoukis, P. and T.P. Labuza, "Applicability of Time Temperature Indicators as Shelf Life Monitors of Food Products," *Journal of Food Science*, 54:783–788, 1989.
(37) Taoukis, P. and T.P. Labuza, "Reliability of Time Temperature Indicators as Food Quality Monitors under Non-isothermal Conditions," *Journal of Food Science,* 54:789–793, 1989.

(38) Malcata, F.X., "The Effect of Internal Temperature Gradients on the Reliability of Surface Mounted Full History Time-Temperature Indicators," *Journal of Food Processing Preservation*, 14:481–487, 1990.
(39) Campden Food and Drink Association, "A Food Industry Specification for Defining Technical Standards and Procedures for Evaluation of Time Temperature Indicators," Technical Manual No. 35, February 1992.
(40) Find SVP, "Time-Temperature Monitoring Products. A Comparative Intelligence Report," New York, N.Y., 175 pages, 1988.
(41) Mistry, V.V. and F.V. Kosikowski, "Use of Time Temperature Indicators as Quality Control Devices for Market Milk," *Journal of Food Protection*, 46(1):52–57, 1983.
(42) Duyvesteyn, W. Integration of the Time-Temperature History Effect on the Shelf Life of Fluid Milk. University of Minnesota: Food Science Thesis, pp. 224, May 1997.

Proposed Regulation:

The proposal recommends the following addition to 21 CFR 101 entitled "Food Labeling".

Part 101-Food Labeling

§101.19 Open Shelf-Life Dating/Open Dating

(a) The purpose of this regulation is to mandate open dating on food products. The open date is intended for use and understanding by both the distributors and the consumers when judging food qualities.

(b) For the purposes of this regulation, the following definitions shall apply.

(1) Open Shelf-Life Date (or Open Date): A date on packaged food to indicate to the consumer when the product was packed or meant to be "used by".
(2) Shelf-life: The amount of time a food product can exist before reaching the end of consumer quality, which is determined by the percentage of consumers that are displeased by the product.
(3) Perishable Food: A food product having an estimated shelf-life of 60 days or less. The definition will exclude unpackaged fresh fruit or vegetable products; however, all foods packaged with reduced oxygen packaging shall be considered a 'perishable food' regardless of its estimated shelf-life.
(4) Semi-Perishable Food: A food product having an estimated shelf-life greater than 60 days but less than 90 days.

(5) Nonperishable Food: A food product having an estimated shelf-life greater than 90 days.
(6) Time-Temperature Integrators (TTIs): Small, physical devices that are placed on the food package to measure the temperature history of a product and indicate a definitive change at the end of shelf-life.
(7) Person: "Person" means an individual, partnership, association, or corporation.
(8) Prepackaged: A food product that is packaged prior to being displayed or offered for retail sale.

(c) For the purpose of this regulation, the following definitions pertaining to open dating shall apply.

(1) Pack or Production Date: The date on which the product was manufactured or put into the final package.
(2) Best-if-used-by: The date where product quality loss reaches a level still generally acceptable but after which it fails to meet the high quality standard.
(3) Use-by: The date determined as the end of shelf-life or the end of useful quality life of a product.
(4) Freeze-by: A date used on food products in conjunction with a use-by date to assist consumers and retailers in terms of product movement.
(5) Closed or Coded Dates: Numbers used by the industry to indicate production lots and dates that are intended for internal industrial use and for recalls.

(d) No person shall sell, offer for sale, or have in their possession with intent to sell, *prepackaged perishable food products* unless they are identified with a "best-if-used-by" date determined by manufacturer, processor, packer, repacker, retailer, or other person who had packaged such food products and displayed in form specified in 21 CFR 101.19 (j). The use of time-temperature integrators are also very highly recommended for use on refrigerated perishable products.

(e) No person shall sell, offer for sale, or have in his possession with intent to sell, *prepackaged semi-perishable food products* unless they are identified with a "best-if-used-by" or "use-by" date determined by manufacturer, processor, packer, repacker, retailer, or other person who had packaged such food products and displayed in the form specified in 21 CFR 101.19 (j).

(f) No person shall sell, offer for sale, or have in his possession with intent to sell, *prepackaged nonperishable food products* unless they are identified with a "best-if-used-by" or "use-by" date determined by manufacturer, processor, packer, repacker, retailer, or other person who had packaged such food products and displayed in the form specified in 21 CFR 101.19 (j).

(g) A manufacturer, processor, repacker, retailer, or other person who *prepackages food products* explained in 21 CFR 101.19 (m) may mark the individual retail packages of such products with a "pack date" or "production date" but shall be subject to the requirements as to form outlined in 21 CFR 101.19 (j).

(h) A manufacturer, processor, repacker, retailer, or other person who prepackages food products may mark the individual retail packages of such products with a closed or code date along with the appropriate open date as described in 21 CFR 101.19 (d) to (g). The purpose of the code or closed date is to further assist production lots of specific products in the event of a product recall or market withdrawal.

(i) No person shall offer for sale in the United States any food product after the expiration of its "best-if-used-by", or "use-by" date unless the food product remains wholesome, is segregated from non-expired products, and is clearly and conspicuously marked as past its open date.

(j) A date shall be displayed with the term "best-if-used-by" or "use-by" in reasonable proximity to the designated date.

(1) Such a date shall consist of the common abbreviation for the calendar month and numerals for the day and year, e.g., Apr. 12, 1999; or numerals for the month, day and year, e.g., 4/12/99, except that:
   (i) Perishable and semi-perishable food products need not have the year identification included in the date, and nonperishable food products need not have the day identification included in the date.
   (ii) Fresh bakery products may be dated with only the day designation, e.g., Monday, or an abbreviation thereof, e.g., Mon.

(2) A date shall be accompanied by disclosure of recommended product storage conditions, if such conditions significantly affect the validity of such a date, such as with refrigerated or frozen foods.

(3) A date and any recommended storage conditions shall be printed, stamped, embossed, perforated, or otherwise shown on the retail package, a label on such package, or a tag attached to such package in

a manner that is easily readable and separate from other information, graphics, or lettering so as to be clearly visible to a prospective purchaser.

(4) If a date and recommended storage conditions do not appear on the principal display panel, the information panel, or on another conspicuous portion of the individual retail package, a statement must appear on the principal display or information panel indicating where such information can be found elsewhere on the package.

(5) An individual prepackaged food product which is not labeled in accordance with provision in Section (a) through (m) shall be deemed "misbranded" pursuant to Section 201 (n).

(k) A person who is responsible for placing an open date on a food product as required or permitted by these regulations, shall estimate the shelf-life of such a product through appropriate shelf-life tests, extensive literature reviews, analysis of distribution data, and/or by common practice. Assuming that the food products are initially prepared as safe and that the product will be stored and handled in proper conditions, the end of shelf-life will represent the end of consumer quality and not a standard of safety. All dating procedures in determining the shelf-life of a food product are subject to review by the federal government. Because certain foods may become unsafe if temperature abused, such products should be accompanied by a TTI and/or the control of temperature distribution be closely regulated.

(l) The following shall apply in regard to record keeping:

(1) A person responsible for estimating the shelf-life of a food product shall keep a record of the method used for the determination of such shelf-life and the corresponding open date in accordance with the premise that there should be facts material to the label claims (Section 201 (n) Food, Drug and Cosmetic Act). A record revision is necessary whenever a factor affecting such date determination is altered. Such record shall be retained for not less than six months after the most recent open date change and be available during normal business hours for examination upon request by the Secretary.

(2) If, after conducting an investigation, the Secretary determines that the date selected is not supported by such records, it may direct the responsible person to change the date in accordance with such findings. Any person aggrieved by such an order shall be afforded the opportunity for hearing, but the order shall not be overturned unless the appellant establishes by clear and convincing evidence that the date originally selected is, in fact, justifiable. The order shall be considered

final unless reversed upon such review; and pending review, the affected product may not be offered for sale, unless the date is modified in accordance with the order of the Department.

(m) The food products listed in 21 CFR 101.19 (m)(1) through (m)(6) are exempt from 21 CFR 101.19 (a) through (l). Any person may apply to the Food and Drug Administration for an exemption from the provisions of 21 CFR 101.19.

(1) Unpackaged fresh fruits and vegetables, or such products which are not packaged in reduced oxygen conditions.
(2) Salt and crystallized refined sugar.
(3) Food products shipped in bulk form for use solely in the manufacture of other foods and not for distribution to the consumer in such bulk form or container.
(4) Individually packaged food products which are prepackaged as components of a larger food item, if the larger food item is identified with a date no later than the corresponding date for any such components.
(5) Food products prepackaged for retail sale with a net weight of less than 1.5 ounces and do not require refrigeration or freezing.
(6) Food products manufactured for sale outside the United States, processed for sale outside of the United States, or stored for sale outside the United States.

(n) 21 CFR 101.19 shall take effect in accordance with the following schedule:

(1) For perishable food, 21 CFR 101.19 (d), the effective date shall be January 1, 2001.
(2) For semi-perishable food, 21 CFR 101.19 (e), the effective date shall be March 1, 2001.
(3) For nonperishable food, 21 CFR 101.19 (f), the effective date shall be January 1, 2002.
(4) For pack or production dates and closed or code dates, 21 CFR 101.19 (g) and (h), the effective date shall be January 1, 2001.

APPENDIX H

ALABAMA SAFE FOODS ACT OF 2000

RULES ADOPTED BY THE AGRICULTURE BOARD
AUGUST 8, 2000

80-1-22-. 31-. 31ER Classifications of Violations for Out-of-Date Class "A" Food and Misbranded or Adulterated Foods

(1) An "Equivalent number" shall be determined using a factor that takes into account the number of packages, case lots, containers or such other packaging and the length of time said products have passed an expiration date. The following factors shall be used to determine an "equivalent number":
   (a) For each package that has passed the expiration date by less than four days the factor would be 0.5.
   (b) For each package that has passed the expiration date by four to seven days the factor would be 0.75.
   (c) For each package that has passed the expiration date by eight to fourteen days the factor would be 1.
   (d) For each package that has passed the expiration date by greater than fourteen and less than 30 days the factor would be 1.5.
   (e) For each package that has passed the expiration date by 30 days or more the factor would be 2.

(2) The equivalent number shall be determined by multiplying the number of out-of-date packages by the appropriate factor as specified in paragraph "(1)" of this rule. An equivalent number shall be determined for each of three primary areas, (Meat, Dairy, Baby Food and Infant Formula), as well as the equivalent number for the aggregate of all three areas. The following criteria shall be used in determining classes of violations for products offered for sale in retail establishments:
   (a) Class I Violations: If each of the three areas have an "equivalent number" less than 20, and the aggregate of all three areas is less than 35, no civil penalty is required. A warning letter may be issued if the "equivalent number" is more than 10 and less than 20 in any of the three categories.
   (b) Class II Violations: If any of the three areas have an "equivalent number" equal to or greater than 20 and less than 40 the violation is considered a Class II violation. If the total of all three areas result in an "equivalent number" equal to or greater than 35 and less than 105 the violation is considered a Class II violation.

(c) Class III Violations: If any of the three areas have an "equivalent number" equal to or greater than 40 and less than 80 the violation is considered a Class III violation. If the total of all three areas result in an "equivalent number" equal to or greater than 105 and less than 130 the violation is considered a Class III violation. Subsequent Class II violations within a one year period shall also result in a Class III violation.

(d) Class IV Violations: If any of the three areas have an "equivalent number" equal to or greater than 80 the violation is considered a Class IV violation. If the total of all three areas result in an "equivalent number" equal to or greater than 130 the violation is considered a Class IV violation. Subsequent Class III violations within a one year period shall also result in a Class IV violation. Products deemed misbranded are considered a Class IV violation. More than ten packaged items offered for sale at retail without proper labeling to include ingredient statements and/or cure ingredients is a Class IV violation and will result in a warning letter. Subsequent labeling violations will result in a penalty administered as a Class IV violation. Obscuring, removing, or extending existing open-date statements shall be considered as misbranding and therefore a Class IV violation.

(e) Class V Violations: Subsequent Class IV violations within a one year period shall result in a Class V violation. Products found to be adulterated are considered a Class V violation. Adulterated products with food safety concerns of an immediate nature will be subject to penalties on the first violation.

(3) The same provisions in paragraph "(1)" of this rule used to determine equivalent numbers shall be used in the assessment of class violations as set forth herein for wholesale packages except that: the unit (package) through which the product is offered for sale shall constitute one package, i.e., case lots, large containers, etc. However, cases of products in storage at a retail facility shall be considered by the individual packages rather than case lots; except that cases of baby food and infant formula in unopened cases shall be considered as one unit. If the equivalent number is exceeded, the penalty shall be assessed in the appropriate class of violation. The following criteria shall be used in establishing class violations for wholesale packages:

(a) Class I Violations: (Wholesale distribution packages) If the out-of-date products have an "equivalent number" less than 5, no civil penalty is required. A warning letter may be issued if the "equivalent number" is more than 1 and less than 5.

(b) Class II Violations: (Wholesale distribution packages) If the out-of-date products have an "equivalent number" equal to or greater than 5 and less than 20 the violation is considered a Class II violation.
(c) Class III Violations: (Wholesale distribution packages) If the out-of-date products have an "equivalent number" equal to or greater than 20 and less than 40 the violation is considered a Class III violation. Subsequent Class II violations within a one year period shall also result in a Class III violation.
(d) Class IV Violations: (Wholesale distribution packages) If the out-of-date products have an "equivalent number" equal to or greater than 40 the violation is considered a Class IV violation. Subsequent Class III violations within a one year period shall also result in a Class IV violation. Products deemed misbranded are considered a Class IV violation. Obscuring, removing, or extending existing open-date statements shall be considered as misbranding and therefore a Class IV violation.
(e) Class V Violations: Subsequent Class IV violations within a one year period shall result in a Class V violation. Products found to be adulterated are considered a Class V violation. Adulterated products with food safety concerns of an immediate nature will be subject to penalties on the first violation.

AUTHOR: Reginald L. Sorrells
STATUTORY AUTHORITY: Act # 2000-320, Regular Session of Legislature, 2000

80-1-22-. 32-.32ER - Open-Date Statement - List of Additional Descriptive Terms

(1) In addition to the terms listed in §20-1-20 (definitions) for (14) open-date-statements, the following list of terms and other terms with similar import, shall also be included and considered as open-date statements:

"For full fresh flavor use by"
"For best quality purchase and use by date shown"
"Use/freeze by"
"Prepare or freeze by"
"For wholesome great taste, serve before date stamped below"
"Best when purchase by date"
"Best if sold by"
"Best used by"
"Product expiration"
"Expiration date"
"Best by"

"Best before"
"Best when purchase by"
"Use before"
"Use by"
"Full freshness until date shown when stored unopened at 40 or below"
"Prepare by"
"Fresh until"
"Use or freeze by"
"Sell or use by"
"Freshness through"

AUTHOR: Reginald L. Sorrells
STATUTORY AUTHORITY: Act # 2000-320, Regular Session of Legislature, 2000

80-1-22-. 33-. 33ER - Definition of Potentially Hazardous Food under Alabama Safe Foods Act of 2000

(1) Potentially Hazardous Food as defined in subparagraph 1-201.10 (B) (61) Food Code, 1999 recommendations of the United States Public Health Service Food and Drug Administration, National Technical Information Service Publication PB 99-115925 and the same is hereby adopted and fully set out herein for reference:
(2) "(61) Potentially Hazardous Food.
    (a) "Potentially hazardous food" means a FOOD that is natural or synthetic and that requires temperature control because it is in a form capable of supporting:
        (i) The rapid and progressive growth of infectious or toxigenic microorganisms;
        (ii) The growth and toxin production of *Clostridium botulinum*; or
        (iii) In raw shell eggs, the growth of *Salmonella enteritidis*.
    (b) "Potentially hazardous food" includes an animal FOOD (a FOOD of animal origin) that is raw or heat-treated; a FOOD of plant origin that is heat-treated or consists of raw seed sprouts; cut melons; or garlic-in-oil mixtures that are not modified in a way that results in mixtures that do not support growth as specified under Subparagraph (a) of this definition.
    (c) "Potentially hazardous food" does not include:
        (i) An air-cooled hard-boil egg with shell intact;
        (ii) A FOOD with an a_w value of 0.85 or less;
        (iii) A FOOD with a pH level of 4.6 or below when measured at 24°C (75°F);

(iv) A FOOD, in an unopened HERMETICALLY SEALED CONTAINER, that is commercially processed to achieve and maintain commercial sterility under conditions of non-refrigerated storage and distribution;

(v) A FOOD for which laboratory evidence demonstrates that the rapid and progressive growth of infectious or toxigenic microorganisms or the growth of *S. enteritidis* in eggs or *C. botulinum* can not occur, such as a FOOD that has an a_w and a pH that are above the levels specified under Subparagraphs (c) (ii) and (iii) of this definition and that may contain a preservative, other barrier to the growth of microorganisms, or a combination of barriers that inhibit the growth of microorganisms; or

(vi) A FOOD that does not support the growth of microorganisms as specified under Subparagraph (a) of this definition even though the FOOD may contain an infectious or toxigenic microorganism or chemical or physical contaminant at a level sufficient to cause illness."

AUTHOR: Reginald L. Sorrells
STATUTORY AUTHORITY: Act # 2000-320, Regular Session of Legislature, 2000

80-1-22-. 34-. 34ER - List of Potentially Hazardous Foods and Exceptions

(1) The following includes, but for enforcement purposes, is not limited to, a list of examples considered to be potentially hazardous foods, and some exceptions.

(a) All Meat products requiring refrigeration to include but not limited to the following:

Fresh or frozen raw meats (poultry, beef, pork, exotics)
Ready to eat cooked meats (fresh or frozen) any specie
Cured meats that require refrigeration
Cooked or heat treated meats that require refrigeration
Processed sausage products that require refrigeration
Fresh refrigerated or frozen chili products
Sandwiches containing meats or other potentially hazardous foods
Seafood
Shellfish
Smoked fish that requires refrigeration
Fish (fresh or frozen)
Fresh or frozen entrees that contain meat
Lunchables (variety packs of meats plus)
Meat salads such as chicken salad, tuna salad, ham salad

(b) Dairy Products requiring refrigeration to include but not limited to the following:

Yogurt
Low fat yogurt
Non fat yogurt
Milk (all forms)
Lactose reduced milk
Acidophilus milk
Chocolate milk
Buttermilk
Cream
Cottage cheese
Sour cream
Salad dressing containing dairy products and requiring refrigeration
Eggnog
Half and half
Cream cheese
Pimento cheese
Pudding that requires refrigeration
Ricotta cheese
Dips with dairy products or meat products that require refrigeration
Butter
Ice cream
Sherbet
Ice milk
Frozen dessert that contains dairy items that requires temp. control
Frozen yogurt
Imitation ice cream
Non fat ice cream
Shake mix
Yogurt mix
Non-dairy mix
Frozen desert mix
Novelty items that contain ice cream type products
Pies that contain eggs or dairy products and require refrigeration
Cheese cakes
Fresh or frozen pastries that contain meat or eggs and require refrigeration
Soft cheese (Brie, Camembert, Teleme)

(c) Egg products requiring refrigeration to include, but not limited to the following:

In shell raw eggs
Egg whites
Egg beaters and similar products
Egg substitutes
Pasta containing eggs that require refrigeration
Eggnog

(d) Items of food that may be excluded from a criteria of potentially hazardous foods include but are not limited to the following:

Chocolate drinks (no milk fat or significant dairy items)
Flavored drinks (no milk fat or significant dairy items)
Biscuits
Margarine
Cookie dough
Jello
Spreads
Cool whip
Shelf stable meat items
Canned meat products that do not require refrigeration
Dry cured items that do not require refrigeration
Jerky type meat items that do not require refrigeration
Fermented meat items that do not require refrigeration
Shelf stable ultra pasteurized milk that doesn't require refrigeration
Bakery desserts (cinnamon rolls etc. that may be refrigerated)

(e) Items of cheese that may be excluded from a criteria of potentially hazardous foods include but are not limited to the following:

Asiago
Cheddar
Gruyere
Parmesan
Reggiano
Romano
Sap Sago
Blue
Brick
Caciocavallo Siciliano
Colby
Edam
Gorgonzala
Gouda
Limburger
Monterey
Monterey Jack
Pasteurized process cheese
Imitation cheese
Provolone
Swiss
Emmentaler
Roquefort

AUTHOR: Reginald L. Sorrells
STATUTORY AUTHORITY: Act # 2000-320, Regular Session of Legislature, 2000

80-1-22-. 35-.35ER - Prohibitions and Exemptions for Altering Open-Date Statements

(1) Packages of potentially hazardous foods bearing an open-date statement are not to be repacked or relabeled or otherwise altered in a manner that would change the open-date statement originally placed on the package. It is not permissible to reprocess products by freezing, slicing, grinding, cubing, dicing, marinating, chopping, or other similar methods unless the original open date statement is maintained on the product label.

(2) In the case of smoking, curing, fully cooking or conversion to a shelf stable product through drying or fermentation of fresh or frozen in-date raw meat products, it may be acceptable to alter the original open-date statement that had been listed on the original product. Provided, however, that in these special situations an appropriate date shall be placed on the product after completion of the process.

80-1-22-. 36- .36 - Exemptions for Shipping Labels and Other Non-Official Labels

Information affixed to containers of meat and poultry food products and case lots of other food products that is not a part of the approved label and is not intended to address quality or condition of the product such as pic labels, routing slips, tracking stickers and similar devices that are intended to provide only shipping data are considered exempt from the dating requirements of the Alabama Safe Foods Act of 2000, except when the commissioner determines such information is subject to be displayed at the retail level and may mislead consumers.

AUTHOR: Reginald L. Sorrells
STATUTORY AUTHORITY: Act # 2000-320, Regular Session of Legislature, 2000

Food Safety Class Violations
Fines & Actions

Class I violations shall be limited to minor or inadvertent violations involving small amounts of out-of-date Class A foods found for which no penalty shall be assessed but a written warning may be given by the department.

Class II violations shall be limited to significant amounts of out-of-date Class A food or such articles which have been out-of-date for significant periods of time. Penalties for Class II violations shall not exceed one hundred dollars ($100).

Class III violations shall be limited to significant amounts of out-of-date Class A foods or significant amounts of such articles which have been out-of-date for significant periods of time; violations in multiple classes; and/or subsequent Class II violations with one year following a previous Class II violation. Penalties for Class III violations shall not exceed one thousand dollars ($1,000).

Class IV violations shall be limited to food deemed misbranded under Section 20-1-25, Code of Alabama 1975; large amounts of out-of-date Class A foods or significant amounts of such articles which have been out-of-date for extended periods of time; violations in multiple classes and/or subsequent Class III violations within one year following a previous Class III violation. Penalties for Class IV violations shall not exceed five thousand dollars ($5,000).

Class V violations shall be limited to food deemed adulterated under Section 20-1-22, Code of Alabama 1975, and/or subsequent Class IV violations. Provided, however, to sustain a penalty under this subsection, with regard to out-of-date Class A foods, there must be a showing by a preponderance of evidence that there were large amounts of out-of-date Class A foods found, or that significant amount of such articles found had been out-of-date for extended periods of time and that the violations were willful, knowing, or intentional. Penalties for Class V violations shall not exceed a maximum penalty amount of ten thousand dollars ($10,000) and may include the revocation of the person's food safety permit.

INDEX

www.ingramcontent.com/pod-product-compliance
Ingram Content Group UK Ltd.
Pitfield, Milton Keynes, MK11 3LW, UK
UKHW020807290926
15470UKWH00003B/1007

9 780917 678530